Beyond the Scan: Using rs-fMRI to Bridge the Gap Between Brain and Behavior

Nimra

TABLE OF CONTENTS

1. INTRODUCTION

The brain is the human organ that causes more fascination and curiosity, mainly because we still poorly understand how this complex organ works. Our eagerness to know about the way the human brain is organized is as old as humanity, and the new technologies have not yet fully addressed the individual's inquietude about mental functions. Nevertheless, progress in neurosciences and particularly in neuroimaging have opened some small windows to the magnificent world of human thinking. Complex functions as perception, emotions, and behavior are ultimately described by neuronal pathways. Once we gain access to understanding those neuronal patterns, either quantitatively and qualitatively, we may better understand the cognitive processes. Knowing how different regions of the human brain interconnect to give rise to a determinate pattern of activation can be useful, not only to understand the neuronal network in normal conditions but also, to understand how diseases can change that normal cognitive network. In the future, different patterns of modulation of brain activity might be used to predict diseases, to anticipate response to treatment and even, eventually, to pick individuals at risk of some condition and act to prevent it. For now, the diagnostic tool is the main purpose of human connectivity and functional mapping studies.

1.1. Objective

The objective of this work is to study brain functional connectivity and functional mapping with resting-state fMRI (rs-fMRI) in healthy volunteers. Although this advanced MRI technique is well described in the literature, the daily clinical application of rs-fMRI is scarce. The majority of studies with patients are carried in research and investigational centers that have funds, equipment, and human resources specifically allocated to these investigations. The main goal of this book is to implement a pipeline to bring rs-fMRI to the clinical practice of the Neuroradiology Department of Centro Hospitalar Universitário do Porto (CHUP). As a national main center in to the Neuroscience field, the actual application of rs-fMRI in CHUP would be of great interest to practitioners and, more importantly, to the patients. Ultimately, the objective is to design a pipeline for the implementation of brain connectivity evaluation by rs-fMRI.

To achieve this main objective specific key-points were delineated:

1) To introduce, develop, and optimize the rs-fMRI data acquisition in a specific MRI scanner of the Neuroradiology Department at CHUP.
2) To pre-process the obtained functional images, using the described methodologies in the literature. This consists of a very important step, once only carefully cleaned data allows for drawing reliable conclusions in further data analysis.
3) To analyze the functional connectivity by rs-fMRI with two different methods: ICA and seed-based analysis, to identify well-established resting-state networks (RSNs).
4) To compare the results of the different processing methods in order to validate them.

1.2. book Outline

The present book is divided into 5 chapters, through which the developed work will be presented and discussed. The present chapter is an introduction to the work, where theoretical background and literature review on relevant studies to the development of the work are presented. Chapter 2 overviews the data used in this study as well as the pre-processing and image analysis performed. Chapter 3 presents the results related to functional connectivity and functional mapping. Chapter 4 discuss the results, its strengths and limitations. Chapter 5 closes the book by presenting the conclusions of the developed work.

1.3. Theoretical background

1.3.1. Brain function and anatomy

The human brain is the most complex organ of the human body, not only functionally but also anatomically, once its intricate anatomy reflects its functional complexity. We briefly review the relevant brain anatomy for this work, to give a background on less familiarized readers.

The human brain is divided into two hemispheres, separated by the longitudinal sulcus, and anatomically connected by the corpus callosum, a median structure of compact white matter fibers. Generally, each hemisphere controls the contralateral side of the body. Despite working together as a whole, not all the functions are represented in both hemispheres. The left hemisphere is usually associated with speech, reasoning, and writing, while the right hemisphere excels in visual perception, spatial ability, and creativity [1]. Each cerebral hemisphere is divided into 6 lobes by sulci and fissures: frontal, parietal, temporal, occipital, insular, and limbic [2].

The frontal lobe is responsible for movement control and for memory, motivation, emotion, reasoning, speech, and language [3]. Functionally, it comprises the primary motor cortex, the supplementary motor area, the premotor cortex, the prefrontal cortex, and Broca's motor speech center (this one unilaterally, on the dominant hemisphere).

The parietal lobe is formed by the primary somatosensory cortex, responsible for the perception of body sensations, and also receive projections of the visual field and auditory spectrums, making it responsible for the integration and comprehension of these complex stimuli [4].

The temporal lobe contains the primary auditory cortex, responsible for receiving and processing auditive information. It is also responsible for some aspects of speech, comprising the Wernicke's sensitive speech center (again, unilaterally on the dominant hemisphere), learning, and memory [5].

The occipital lobe includes the primary visual cortex, responsible for receiving and processing visual information and interpretation of the visual stimuli [6].

The insular cortex receives input from its connections to the primary and secondary somatosensory cortex, orbitofrontal cortex, and inferior parietal lobule. Its role is not fully understood, but it is thought to modulate feelings and emotions, recognition of fine touch, auditory impulses, and some language connections [7].

The limbic structures comprise the hippocampus, amygdala, and olfactory bulb, being responsible for the regulation of emotions, motivation, behavior, and memory [8].

In addition to the hemisphere lobes, basal ganglia and thalami are gray matter structures in deep location, that are responsible for the control and planning of stereotyped movements, regulation of posture and muscle tone adjustment. They influence motor activity by sending impulses through the thalamus to the cerebral motor and premotor cortex [9]. The thalami work as an integration center for several stimuli, being part of several connection circuits.

In the rostral part of the brain is the cerebellum that intimately connects with the motor cortex to coordinate movements. It is crucial to motor learning and reflex modification [10].

1.3.2. Functional Magnetic Resonance Imaging (fMRI)

1.3.2.1. Principles of Magnetic Resonance Signal Generation

Magnetic Resonance Image (MRI) is an imaging technique based on the nuclear magnetic resonance (NMR) phenomenon described independently by Purcell et al and Bloch et al in 1946 [11]. Only many years later, in 1973, the NMR was applied to the image of the human body when Lauterbur and Mansfield [12] first described the use of the MRI. It is a non-invasive imaging modality that has experienced rapid growth over the last decades, still improving every day. MRI has high contrast and spatial resolution and does not require the use of ionizing radiation, being one of the best and most clinically useful techniques in the study of the human brain.

All atoms with an odd number of protons and/or neutrons possess a nuclear spin angular momentum $(\vec{J})$, often just called spin. For human imaging purposes, the MRI signal comes predominantly from the excitation of the hydrogen $(_{1}^{1}H)$ magnetic moment, as it is the most abundant MR active nucleus in biological tissues [13].

A nucleus with a non-zero angular momentum $(\vec{J} \neq 0)$, exhibits a nuclear magnetic moment, $\vec{\mu}$, called nuclear magnetic dipole or magnetic moment. These two are related to each other by one of the basics equations of particles' physics [14]:

$$\vec{\mu} = \gamma \vec{J} \tag{1}$$

where γ is the gyromagnetic ratio, which value depends on the atomic species. For hydrogen, $\gamma = 267.52 \times 10^{6}\ rad/s/T$. This gyromagnetic ratio can be positive or negative: when positive it means that the magnetic moment $(\vec{\mu})$ is parallel to the angular moment $(\vec{J})$, while when negative the magnetic moment $(\vec{\mu})$ has opposite direction to the angular moment $(\vec{J})$.

To define the magnetic moment (μ), one must know its magnitude and orientation. The magnitude is given by the following equation, according to the theory of quantum mechanics [15]:

$$\mu = \gamma \hbar \sqrt{I(I+1)} \tag{2}$$

where $\hbar$ relates to the Plank's constant h ($\hbar = h/2\pi$) and I is the nuclear spin quantum number. Nuclei with an even mass number and even charge number have null spin ($I = 0$) while nuclei with an odd charge number and/or odd mass number have nonzero spin ($I \neq 0$). In biological tissues, the orientation/direction of $\vec{\mu}$ depends on the existence of an external magnetic field. In the absence of the external magnetic field, the $\vec{\mu}$ direction will be random (as a result of random thermal motion of the nuclei), resulting in null total magnetization. On the contrary, after turning on a strong external magnetic field ($\vec{B}_0$), the nuclei magnetic momenta ($\vec{\mu}$) tend to align with the direction of the magnetic field (Figure 1). When under a magnetic field, the spins will align with the direction of the field and will precess around it (Figure 2). This means that the nuclei magnetic momenta will precess around $\vec{B}_0$. Considering a xyz coordinate system, the applied external magnetic field $\vec{B}_0$ is assumed to be along z direction.

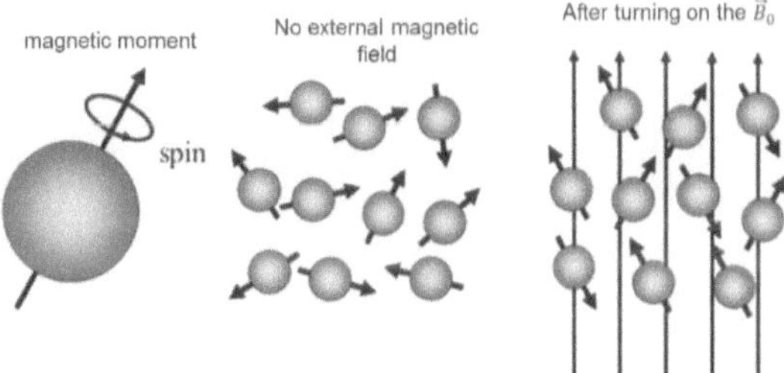

Figure 1: Representation of the magnetic moment of a spin, the random orientation of the spins in the absence of an external magnetic field and the alignment of the spins when under a magnetic field. (Adapted from [16])

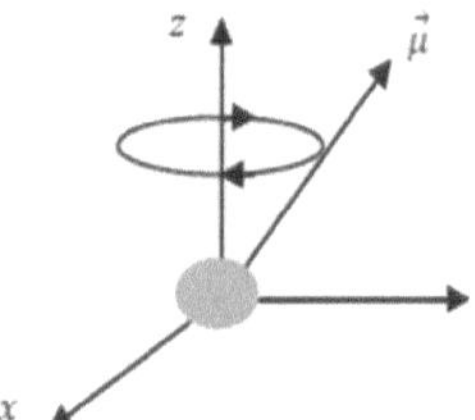

Figure 2: Precession of a nuclear spin around the main magnetic field $\vec{B}_0$. (Adapted from [17])

As so, the magnetic moment will also be precessing along z and can, therefore, be represented by μ_z. Its magnitude is given by the following equation [17]:

$$\mu_z = \gamma m_I \hbar \tag{3}$$

where m_I is the magnetic quantum number. For nuclei with nuclear spin of $1/2$, m_I takes the values of $\pm 1/2$, meaning that $\vec{\mu}$ will have two possible orientations in relation to the applied external magnetic field: parallel orientation ($\uparrow$) if positive m_I or anti-parallel ($\downarrow$) if negative m_I. Depending on the orientation of the spins, they will have different energy levels as result of the interaction with the external magnetic field. (Figure 3).

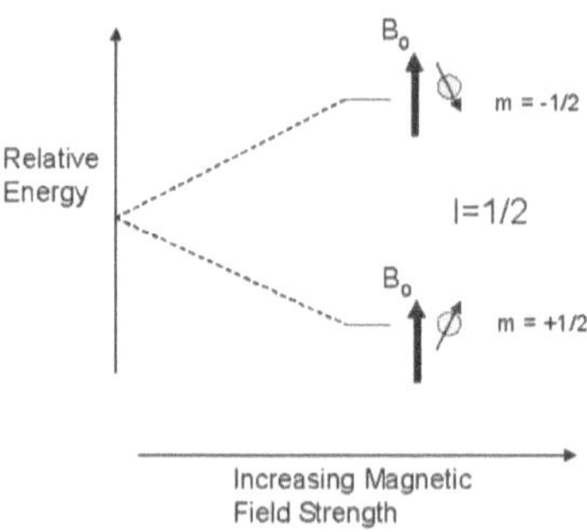

Figure 3: Representation of parallel and anti-parallel orientation. Spin states which are oriented parallel to the external field are lower in energy than in spin states whose orientations oppose the external. (Adapted from [18])

Based on quantum theory, the spin's energy under a uniform magnetic field $\vec{B}_0$ is given by [17]:

$$E = -\vec{\mu}.\vec{B}_0 = -\mu_z.B_0 = -\gamma\hbar m_I.B_0 \qquad (4)$$

Meaning that, for each different spin, the energy can be [17]:

$$E\left(m_I = {}^1\!/_2\right) = -\gamma\hbar B_0/2 \qquad (5.1)$$

$$E\left(m_I = -{}^1\!/_2\right) = \gamma\hbar B_0/2 \qquad (5.2)$$

Interpreting these equations, we can understand that the parallel orientation state $(m_I = 1/2)$ is the state of lower energy while the anti-parallel orientation state $(m_I = -1/2)$ is the state of higher energy. The energy difference between both states (often designated Zeeman effect) is calculated as follow [17]:

$$\Delta E = E_\downarrow - E_\uparrow = \gamma\hbar B_0 \qquad (6)$$

The number of spins on each state (population of the different spin states - n) is related to this energy difference by the Boltzmann distribution [17]:

$$\frac{n_\uparrow}{n_\downarrow} = e^{\frac{\Delta E}{KT}} \qquad (7)$$

where T is the absolute temperature (in kelvin - K) of the spin system and K is the Boltzmann constant ($K \approx 1{,}38x10^{-23} J/K$).

All spins contribute to the global magnetization, and the sum of each individual microscopic magnetic momentum inside a volume gives rise to the net magnetization vector $\boldsymbol{M}$. In the absence of the static external magnetic field, and knowing that all magnetic momenta of individual spins are independent, the net magnetization $\boldsymbol{M}$ will be null given the random orientation of the spins. Assuming that the external magnetic field $\vec{B}_0$ is along the z direction, as we have seen before, the magnectic momentum of each spin will also align and precess along the z direction, which will be the direction of the net magnetization vector $\boldsymbol{M}$. The magnetization vector will be maximum when all spins are lined up with the main magnetic field. This vector is described by three components: M_z, corresponding to the longitudinal magnetization, and M_x and M_y, both corresponding to the transversal magnetization, on xy plane; they are commonly combined and represented by M_{xy}, known as transverse magnetization. The precessing movement of the net magnetization $\boldsymbol{M}$ is analogous to the one of the individual spins. In that way, if we again consider a xyz coordinate system, the $\boldsymbol{M}$ magnetization will be precessing around the z-direction (Figure 4).

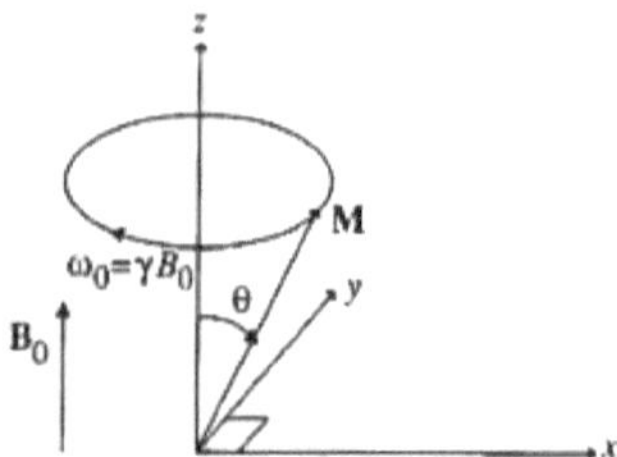

Figure 4: Precession of the net magnetization $\boldsymbol{M}$ around $\vec{B}_0$, along z-direction (Adapted from [19])

As we have seen, the longitudinal magnetization is maximum at equilibrium, when the spins are more likely to assume the low energy state (parallel) than the high energy state (anti-parallel). The difference between the spins population is very small, but is sufficient to give rise to an observable macroscopic magnetization along the longitudinal component M_z, known as equilibrium magnetization $\boldsymbol{M}_0$ [20]. Since the spins do not rotate in phase, the sum of all the microscopic transverse magnetization of the spins results in a null macroscopic transverse magnetization (Figure 5).

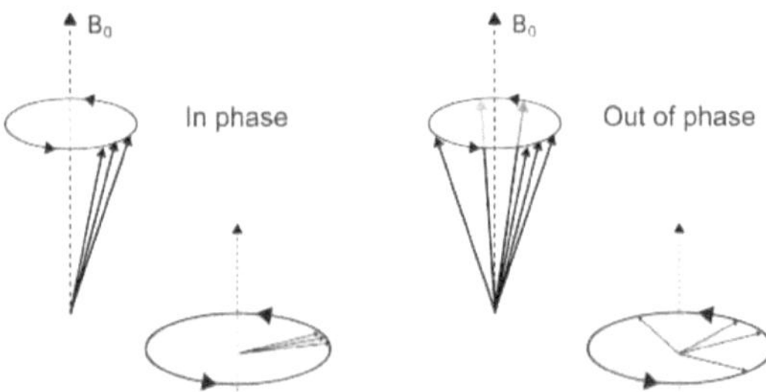

Figure 5: Illustration of the dephasing of preccessing spins in the xy plane, making the transverse magnetization null. (Adapted from [21])

To generate MRI signal (magnetic resonance signals emitted by the nuclei when returning from an excited state to the equilibrium), the magnetization have to change in time and the magnetic momenta must have to be excited by a magnetic component $\vec{B}_1$ of the radiofrequency circularly polarized wave (RF-wave) in a process called spin excitation. If the frequency of the exciting RF-pulse is equal to the spontaneous resonance frequency of the spin system (Larmor frequency - ω_0), the excitation process is in resonance. The Larmor frequency of the system, also called precession frequency, depends on the atomic species, and is given by [17]:

$$\omega_0 = \gamma B_0 \tag{8}$$

The Larmor frequency is directly proportional to the external applied magnetic field B_0 with the gyromagnetic ratio (γ) of the atom in the system, being the proportional constant.

When the RF-pulse is applied perpendicularly to the $\vec{B}_0$, there is a torque that rotates the magnetization M, giving rise to the transversal components of the magnetization, making M_x and M_y measurable. In the simplest case, assuming the rotating frame, the circularly polarized RF-pulse can be described by a rectangular envelope with duration Δt, where the rotation angle θ is given by [17]:

$$\theta = B_1 \, \Delta t \tag{9}$$

Given this, as a result of the excitation, the bulk magnetization is tipped away from the direction of $\vec{B}_0$ reducing the longitudinal magnetization and creating a measurable transverse magnetization [20]. If the RF-pulse rotates the net magnetization into the transverse plane, it is called a 90° RF pulse and the transverse magnetization will be maximum.

The analysis of the net magnetization movement after an RF-pulse, in the stationary frame (laboratory frame) is complex. Considering the stationary frame xyz, where $\vec{B}_0$ is along z-direction, the application of the $\vec{B}_1$, let's say along the x direction, will tip the magnetization to the xy plane, describing a nutation movement (Figure 6).

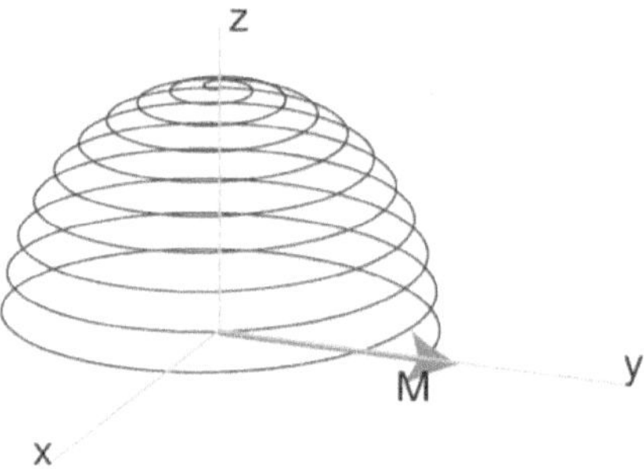

Figure 6: The magnetization M along the z-axis, is seen in the stationary frame to follow a nutation around the z-axis due to the static magnetic field and also around x-direction due to a RF pulse perpendicular to the z-direction. (Adapted from [22])

To better describe the magnetization movement is preferable to consider a rotating frame $x'y'z'$, where z' conincide with z, having the same direction as $\vec{B_0}$. This way, x' and y' axis rotate around z' with a precessing frequency equals to Larmor frequency - ω_0. In the rotating frame, if we apply the $\vec{B_1}$ magnetic field in the x' direction, the magnetization movement can be interpreted as a direct rotation from z' to $x'y'$ plane, with a rotatory movement around $\vec{B_1}$ [17]. This way, the $B_1(t)$ will appear static in the rotating frame (Figure 7).

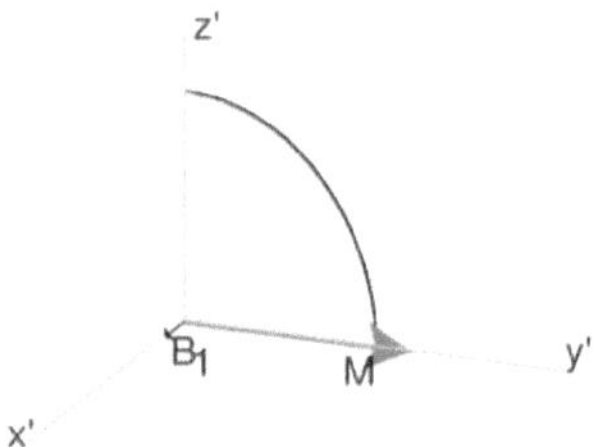

Figure 7: Representation of the rotating frame on resonance with the precessing spins, so that no field exists along z. (Adapted from [22])

Relaxation

After removing the RF-pulse, the net magnetization will evolves towards the thermal equilibrium value under the influence of the external magnetic field $\vec{B}_0$; this means that the transverse magnetization component goes to zero – transversal relaxation, while the longitudinal component recovers the equilibrium value – longitudinal relaxation.

Longitudinal relaxation time T1

When the magnetization starts to grow back in the longitudinal direction, we have longitudinal or T1 relaxation (Figure 8). T1 is a time constant representing the time interval taken to the longitudinal magnetization to return to 63% of its original value (assuming a 90° RF pulse) and is also called spin-lattice relaxation (Figure 9). The recovery of the longitudinal magnetization can be expressed mathematically by [20]:

$$M_z(t) = M_0 \left(1 - e^{-\frac{t}{T1}}\right) + M_z(0)e^{-\frac{t}{T1}} \tag{10}$$

where M_0 is the equilibrium value of the longitudinal magnetization and $M_z(0)$ is the longitudinal magnetization right after the RF-pulse is applied. By the equation, we understand that the longitudinal magnetization grows exponentially with a time constant T1.

The T1 time constant depends on several factors namely the magnetic field strength, the materials present in the medium and its physical characteristics, as well as the physical properties of the nuclei that originated the signal. T1 can be used to distinguish different tissues.

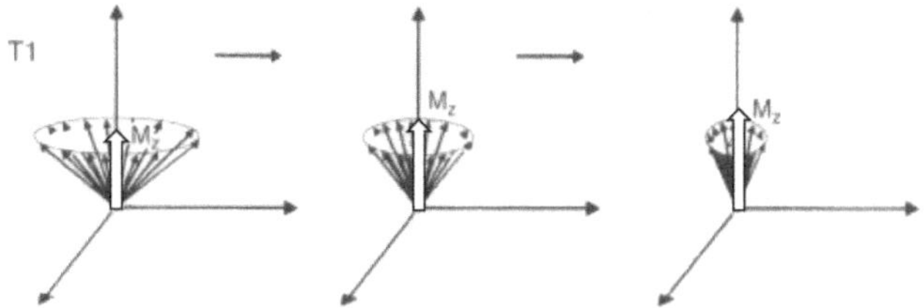

Figure 8: Diagram of T1 (spin-lattice, longitudinal) relaxation after a 90° nutation pulse. (Adapted from [16])

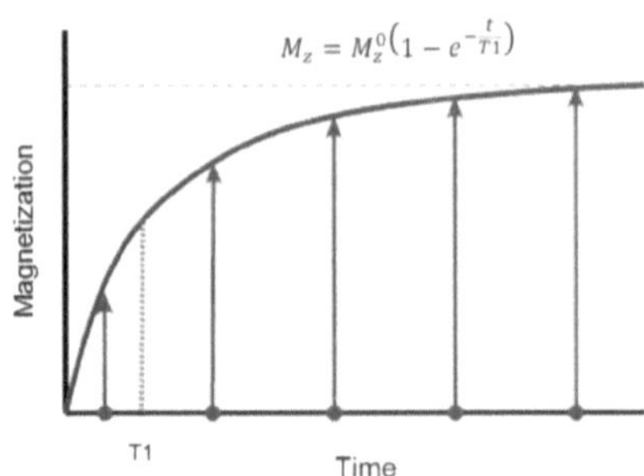

Figure 9: Longitudinal relaxation is modeled as exponential growth curve with time constant $T1$, assuming a 90° RF-pulse. Mz reaches 63% of its maximum value (M_z^0) at $t = T1$ and is very close to maximal at $t = 5 \times T1$. (Adapted from [23])

Transverse relaxation time T2

The transverse relaxation occurs with the dephasing of the individual spins on the xy plane, caused by a magnetic field variation. The loss of phase coherence will traduce into destruction of the transverse magnetization (Figure 10). The described process, also denominated spin-spin interaction, is determined by the T2 time constant, that represents the time interval taken to the transverse magnetization to fall to 37% of its original value (assuming a 90° RF pulse) (Figure 11). The decay of the transverse magnetization can be expressed mathematically by [20]:

$$M_{xy}(t) = M_{xy}(0)e^{-\frac{t}{T2}} \tag{11}$$

where $M_{xy}(0)$ represents the transverse magnetization right after the RF-pulse is applied (thus corresponding to the maximum transverse magnetization). Similar to the T1, T2 time constant depends on several parameters including the magnetic field strength, the molecular structure of the medium and its physical characteristics. As for T1, different tissues have different T2 values.

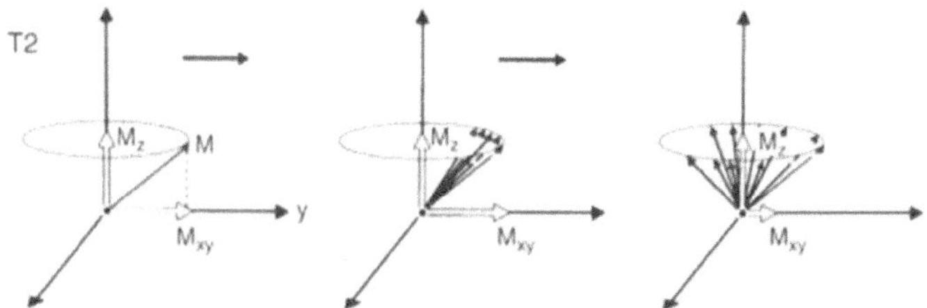

Figure 10: Diagram of T2 (spin-spin, transverse) relaxation after a 90° nutation pulse. (Adapted from [16])

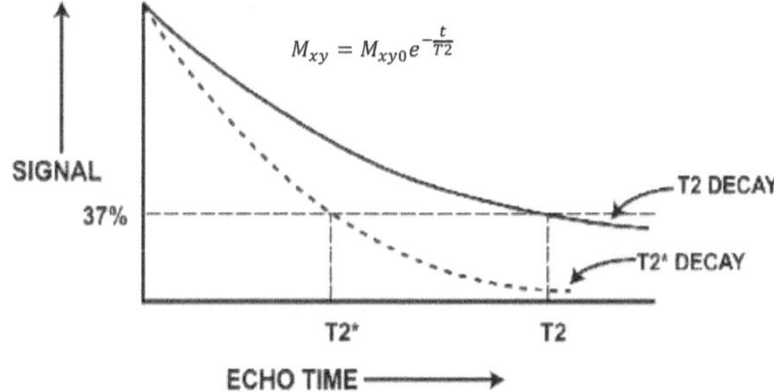

Figure 11: Transverse relaxation. Graphic representation of T2 and T2* relaxation curves (T2* is shorter than T2). Mxy reaches 37% of its maximum at $t = T2$. (Adapted from [23])

Transverse relaxation time T2*

The transverse dephase of the spins occurs due to inhomogenities in the magnetic field, spin-spin interactions, magnetic susceptibility and also chemical shift artefacts. These local inhomogeneities cause a rapid dephasing, faster than it would be if the spins where submitted just to the magnetic field. This results in acceleration of the FID signal, characterized by a new relaxation time T2*. This interaction can be mathematically translated by the following expression [24]:

$$\frac{1}{T2^*} = \frac{1}{T2} + \frac{1}{T2_{inhomogeneties}} \tag{12}$$

Free Induction Decay

The initially coherent transverse components of M dephase as a result of both magnetic field inhomogeneities and intrinsic T2 mechanisms, incorporated in the concept of T2*-decay described earlier. During the relaxation process protons will emit electromagnetic energy (nuclear magnetic signal) also known as the free induction decay (FID) signal [25]. The FID oscillates at the Larmor frequency but is damped by the T2* decay (Figure 12). The resulting FID signal is a damped sine wave of the following form [20]:

$$[\sin \omega_0 t]e^{-\frac{t}{T2^*}} \tag{13}$$

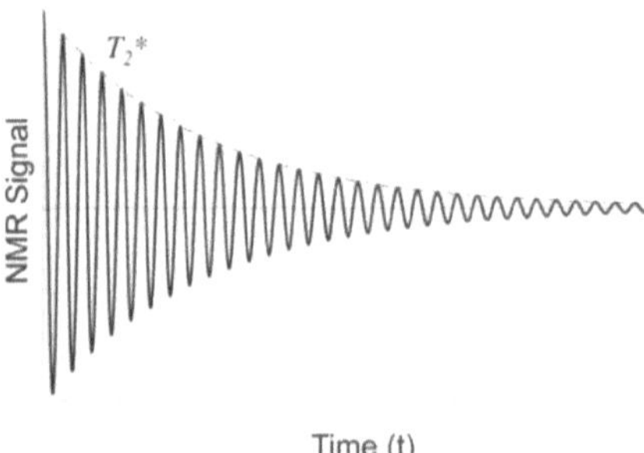

Figure 12: Free Induction Decay (FID) nuclear magnetic resonance signal (Adapted from [23])

Spatial localization of MR signal

Spatial localization is based on magnetic field gradients, applied successively along different axes. Magnetic gradient causes the field strength to vary linearly with the distance. These gradients are employed for slice selection, phase encoding and frequency encoding [26]. The different gradients used to perform spatial localization have identical properties but are applied at distinct moments and in different directions. This way, after applying all three gradients, we can precisely localize the origin of one specific MR signal from a determinate region of the human body.

<u>1.3.2.2. Basic Principles of fMRI</u>

fMRI is an advanced MRI technique, first described in the 1990s, that opened the window to the study of neuronal activity [27]. This is a safe, non-invasive method to study brain activation through measurements of changes in blood oxygenation during a specific task performed by the human brain. The concept of fMRI was first introduced by Belliveau et al [28] in 1991, when the authors used dynamic susceptibility contrast to measure the changes in cerebral blood volume following neural activation caused by visual stimulation, aiming to create a functional magnetic resonance map of human task activation.

fMRI has a fairly good spatial resolution and good temporal resolution which, along with the fact that it doesn't need the injection of the radioactive agent (being noninvase), constitute the main advantages facing Positron Emission Tomography (PET) studies, the other method available to evaluate brain activity [29].

BOLD Phenomenon

Neurons do not contain any internal source of energy, either in form of oxygen or glucose. Therefore, when activated, they need to be supplied with more energy by the circumjacent capillaries that provide them with the required oxygen, through a process called hemodynamic response [30]. This hemodynamic response in consequence of the neural activation is known as the neurovascular coupling. The neurovascular coupling is the basic phenomenon behind the fMRI since it allows us to infer the neuronal activity based on the increase of the blood supply to that brain tissue being activated. The most common method to perform fMRI is based on the Blood Oxygen Level Dependent (BOLD) phenomenon that has a measurable effect on the MR signal intensity of the cerebral tissue being analyzed. The BOLD signal was first described by Ogawa in 1990 [31]. One interesting aspect about the neurovascular coupling and the hemodynamic response is that the increase in the neuronal activity increases the metabolic demand, which in turn recruit blood flow in more quantity than the one needed for blood supply, resulting in excess of blood, and therefore in oxygen, in the activated region (Figure 13). The BOLD signal measured depends on this difference in oxygenation between metabolic active tissues and resting tissues [29].

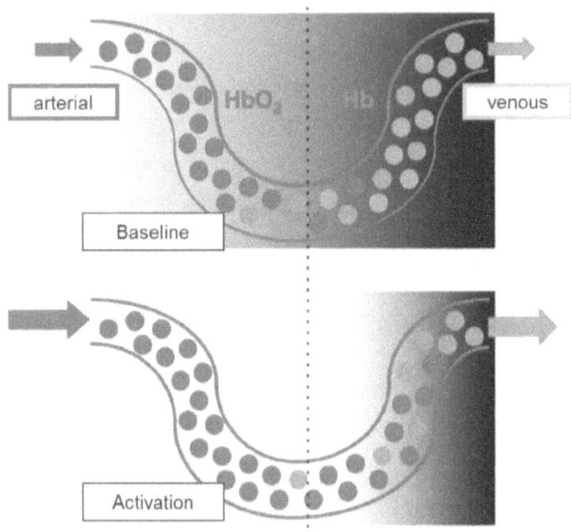

Figure 13: Hemodynamic changes in blood oxygen level. During activation there is increased blood flow and blood volume that cause rising in the oxyhemoglobin/deoxyhemoglobin ratio, thus decreasing the deoxyhemoglobin in the tissue. (Adapted from [32])

Decades before Ogawa has described the BOLD signal, Pauling and Coryell, in 1936, discovered that the magnetization level of hemoglobin depended on its levels of oxygenation [33]. When deoxyhemoglobin was placed under a magnetic field, the authors observed that it was strongly attracted by the field, unlike oxyhemoglobin. The BOLD contrast is therefore based on the changes in the ratio between oxyhemoglobin (Hb) and deoxyhemoglobin (dHb), the latter being the metabolite contributing to the activation-induced susceptibility changes and consequent local signal decrease. As the relative quantity of deoxyhemoglobin decreases in the active areas, the MRI signal increases when compared to the surrounding tissues.

This BOLD phenomenon was first studied as a response to a stimulus or task inducing brain activity (increase in synaptic activity and electric conduction between neurons). This activity increases the blood flow – cerebral blood flux (CBF) and consequently the blood volume – cerebral blood volume (CBV). The brain activity leads to an increase in the cerebral metabolic rate of oxygen ($CMRO_2$) as well as in the cerebral metabolic rate of glucose (CMR_{Glu}) [34]. This induced vascular hemodynamic response will produce changes in the constituents of the brain tissue being activated, which will give rise to changes in the microstructural magnetic field of that region, leading to changes in the MRI signal. In Figure 14, we can see how the changes in the biophysical parameters will influence the MRI signal.

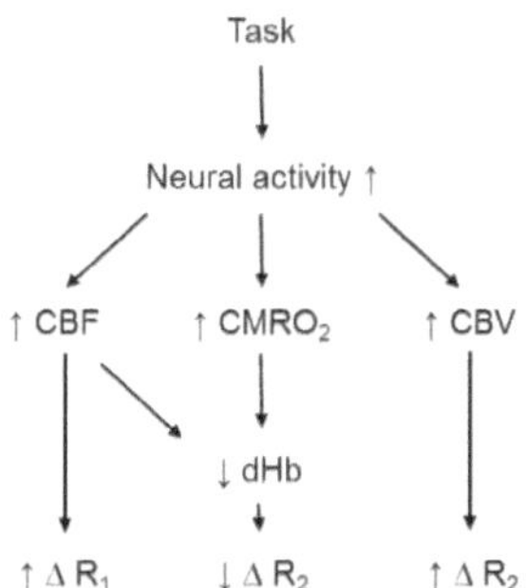

Figure 14: Schematic representation of MRI signal changes induced by brain stimulation. Task/stimulation increases neural activity and increases metabolic (CMRO$_2$) and vascular responses (CBF and CBV). Increase in CBF enhances venous oxygenation level, whereas increase in CMRO$_2$ decreases venous oxygenation level. Because the increase in CBF exceeds the increase in CMRO$_2$, venous oxygenation level increases. These vascular and metabolic changes will modulate the local magnetic field resulting in changes of biophysical parameters. Increases in CBF and CBV increase R$_1$ (= 1/T1) and R$_2$ (= 1/T2), respectively, whereas decrease in dHb contents reduces R$_2$. Changes in R$_1$ and R$_2$ will therefore change the T1 and T2 that will in turn influence the MRI signal. (Adapted from [35])

Assuming an arterial oxygen saturation of 1.0, the relative change of venous oxygenation level (Y) can be determined from the relative changes of both CBF and CMRO$_2$ as described by Ogawa [36]:

$$\frac{\Delta Y}{1-Y} = 1 - \frac{(\Delta CMR_{O_2}/CMR_{O_2} + 1)}{(\Delta CBF/CBF + 1)} \tag{14}$$

We should be aware that we are measuring neuronal activity indirectly and that the neural activity is much faster than the hemodynamic response that follows: the neuronal activity can last milliseconds while the vascular response will take few seconds to reach it maximum (usually around 5 seconds). The relationship between oxygenation and blood flow change is linear at low CBF changes [37] but can be nonlinear at very high CBF changes. It is also important to note that the BOLD response acts as a linear time-invariant system and the linearity of the BOLD is crucial to the analysis of the results allowing the use of General Linear Models (GLM), which will be discussed further on.

Acquisition sequence

The most used image acquisition sequence for fMRI purposes is the echo-planar image (EPI). In the need for imaging the physiological markers of transient events and processes, the MR signal needs to be reconstructed to the entire image in the shortest time possible, but, at the same time, lasting sufficiently long enough for BOLD-inducing factors influencing the MR signal to develop [35]. The demand for higher temporal resolution requires more rapid MR signal sampling. In order to study the changes in the hemodynamic response of brain tissue, we must use a time and physiologic sensitive sequence.

This is accomplished with shorter acquisition times, which is in turn attained with sequences that cover the entire k-space after a single excitation pulse. The group of imaging methods that covers the all k-space in a single passing shot, or in a series of multiple shots, constitutes the Echo Planar Imaging (EPI) techniques. EPI pulse sequences are spatial encoding schemes that use the same echo-formation mechanisms as spin or gradient-echo methods (allowing different images such as Spin-echo EPI, Gradient-echo EPI and Inversion-Recovery EPI), with the only difference that they are very quick to traverse the entire k-space, thanks to several rephasing gradients for the same exciting pulse [38]. From all echo-planar images, gradient-echo EPI is the most frequently used technique [39].

Echo-planar imaging has some advantages compared to conventional MRI, namely the shorter time of acquisition (allowing better temporal resolution), the decreased sensivity to motion artifacts and the capacity to image rapid physiologic processes of the human body [39]. However, EPI presents some disadvantages, the most important being the artifacts caused by inhomogeneities in the magnetic field and geometric distortions in the boundaries between brain tissues, bone, and air-filled cavities.

1.3.2.3. Resting-State fMRI

The fMRI developed over the years, and in 1995, Biswal [40] performed the first study on human brains to assess the functional connectivity in the motor cortex of the resting human brain. These authors found correlations of low frequency between different areas of the brain that brought attention to the connectivity within the brain. With this discovery, several studies started to focus on this issue and there was a boom in publications about resting-state connectivity.

Resting-state fMRI evaluates the spontaneous fluctuations in brain activity without stimulus or task paradigms. The basic principle is to identify synchronous alterations on the BOLD signal of different brain regions. Because rs-fMRI does not need the compliance of the subjects to adhere to some task, it can be applied to populations incapable of performing task-based functional MR imaging such as children, subjects with dementia, and patients with reduced consciousness (coma or sedation) [41]. Moreover, the rs-fMRI analysis is not restricted to one single cognitive domain like the task-based functional MR imaging, allowing the multi-domain analysis simultaneously. In addition, rs-fMRI enables the evaluation of functional connectivity networks and their inter-relationships all across the brain. All these advantages turn rs-fMRI very attractive to the clinical practice application.

1.3.3. Brain connectivity

Although brain anatomy has been extensively studied and is almost perfectly known, brain functionality and connectivity is still an evolving field. For many decades now, there has been a debate on whether specific mental functions are located in specific cerebral areas or instead they are more diffusely distributed on the brain. Nowadays, most neuroscientists agree that there is at least some degree of localization of mental function, but the function of each of these regions must be integrated in order to achieve coherent mental function and behavior [29]. Brain connectivity datasets comprise networks of brain regions connected by anatomical tracts or by functional associations [42]. Neuroimaging research must take functional integration seriously to fully explain brain function [43].

Functional connectivity

The correlation of the activity of different and spatially distant regions of the brain reflects the brain functional connectivity. These activations must be synchronous and can arise for numerous reasons. *Effective connectivity* describes the direct influence of one region on another [44]. However, there are several forms of indirect connectivity: (i) one region may exert its influence on another region under the mediation of a third region or (ii) there can be two separate regions receiving inputs from one same area, being thus activated simultaneous, even if they do not influence each other alone. These concepts may arise concerns on whether the perceived functional association reflects true direct causal influence between two anatomically distinct regions.

It is known that "the resting human brain represents only 2% of total body mass but consumes 20% of the body's energy, most of which is used to support ongoing neuronal signaling. Task-related increases in neuronal metabolism are usually small (<5%) when compared with this large resting energy consumption" as stated by Fox et al [45]. In this way, it is legit to think that the connectivity of the brain in resting-state is the key to fully understand how the human brain works, once it is the process where the human brain expends more energy.

In this sense, rs-fMRI can be used to study the brain's functional connectivity and how different areas of the brain interconnect and activate synchronously with each other to give rise to a determinate functionality. The study of brain functional connectivity also allows the construction of functional maps, which are representations of how some areas of the brain are activated together and act as a whole to allow a complex task.

These functional maps have been studied by several researchers in big databases of healthy controls and diseased individuals, allowing to collect data about the patterns of brain connections in the human brain that are more reproducible. There are about 20 functional patterns of brain connections that can be acquired regardless of the acquisition method and across multiple different individuals. These connectivity patterns are called resting-state networks, each consisting of temporally synchronized structures. Over the years there were several RSNs that were found more inconsistently and there is still some debate as to include them as RSNs. Nevertheless, we will be focusing on the 10 more consensual RSNs that were described and summarized by *Smith et al* [46]. These 10 RSNs include the visual medial, visual occipital and visual lateral area; default mode network; cerebellum; sensorimotor area; auditory area; executive control; right frontoparietal and left frontoparietal as depicted in Figure 15.

The anatomic distribution and the specific function of the most commonly described RSNs in the literature are resumed on Table 1. In this Table, we included the RSNs described by *Smith et al*, but also the Salience and Dorsal Attention networks described more recently [47].

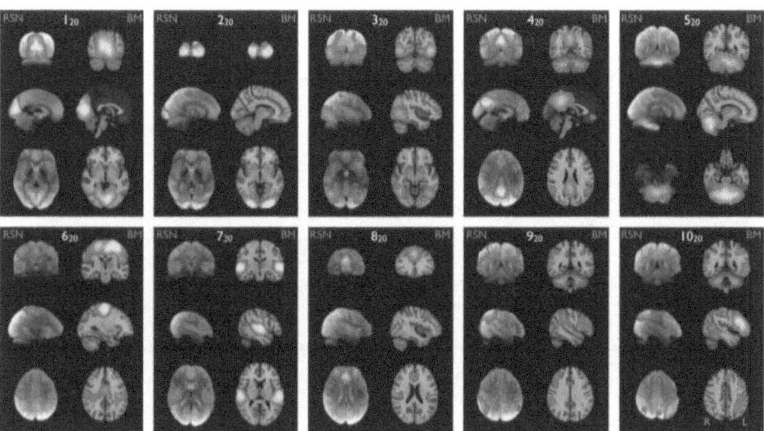

Figure 15: Image from *Smith et al* [46]. Depiction of the 10 RSNs in 3 orthogonal slices, both on fMRI images (left column for each) and on MNI standard space (right column for each). From 1 to 10 they represent respectively: medial visual area (1), occipital visual area (2), lateral visual area (3), default mode network (4), cerebellum (5), sensorimotor (6), auditory (7), executive control (8), right frontoparietal (9) and left frontoparietal (10).

Table 1: Anatomy and function of the RSNs.

Resting-state networks	Anatomical structures	Function
DMN	Precuneus, posterior cingulate, bilateral inferior-lateral-parietal and ventromedial frontal córtex	Episodic memory
Visual networks	Medial, occipital, lateral areas	Field of cognition-language and cognition-space
Sensorimotor network	Primary sensorimotor cortex, suplemmentary motor area and secondary motor cortex.	Motor functions
Executive control	Anterior cingulate and paracingulate	Working memory
Auditory network	Superior temporal gyrus and posterior insular	Execution of speech and auditory perception
Frontoparietal networks	Many frontoparietal areas	Language and cognition

		Social behaviour, self-awareness through the processing of sensory, emotional, and cognitive information
Salience network	Dorsal anterior cingulate cortex and anterior insula	Social behaviour, self-awareness through the processing of sensory, emotional, and cognitive information
Dorsal-attention network	Superior parietal and superior frontal areas, including intraparietal sulcus	Focused attention and orientation
Cerebellum	Cerebellum	Motor learning, fine movement and equilibrium

Adapted from Smith et al [46] and Barkhof et al [41].

The default mode network (DMN) is the most studied and the easiest identifiable pattern. It is a resting network, meaning that is stronger and more active at rest, decreasing during task-based fMRI stimulus. DMN is known to be a task-negative network. On the contrary, task-positive networks include visual, auditory, executive control, sensorimotor, frontoparietal, and cerebellum networks. The two groups are in balance and work in opposite directions, which means that when task-positive networks are active the DMN decreases its activation, and vice-versa.

Variations on the strength and activation of these different networks can be found in different situations in healthy individuals (reflecting physiologic differences) and also on different pathologic conditions. Sleep is one of the states that can make these networks vary, namely decreasing the activation of DMN [48]. Also, the administration of some drugs has interference with the activation of the networks [49, 50]. Finally, it is now well established that aging has an important role in modulating the RSNs, particularly by decreasing DMN activity [51]. Besides these changes in physiologic/pharmacologic alterations, RSNs are also modified in several different pathologies, with neurodegenerative diseases being the most studied field [52].

In fact, several authors have described that Alzheimer's disease and other neurodegenerative diseases present a decrease in activation of the DMN comparing to age-matched controls [53, 54]. The capability to understand these differences has clinical importance since these alterations tend to appear earlier than the structural changes and therefore can potentially be used as disease markers.

Nevertheless, the application of the brain connectivity study in the clinical setting is still investigational, with few centers applying these image method to diagnose and orientate individual patients.

The rapid pace of development and the interdisciplinary nature of the diverse fields that use and collaborate to fMRI data presents an enormous challenge to researchers [55]. The ability to evolve rs-fMRI requires strong collaborative teams with expertise in a number of disciplines, including psychology, neuroanatomy, neurophysiology, physics, biomedical engineering, signal processing, and statistics. True interdisciplinary collaboration is extremely challenging, as all members of the research team must know enough about the other disciplines to be able to talk intelligently with experts in each discipline [56].

1.3.4. fMRI data analysis

In order to analyze the fMRI data, a software package is usually necessary. In the early days of fMRI every single lab/department wanting to perform data analysis needed to have its own software package. Nowadays, there are several software packages available for fMRI data analysis (Table 2). These are full-fledged analysis suites, able to perform all aspects of the analysis of an fMRI study and that can be used by anyone.

Table 2: Overview of major fMRI software packages.

Package	Developer	Platform
SPM	University College of London	MATLAB
FSL	Oxford University	Linux, macOS, Windows*
AFNI	NIMH	Linux, macOS
Brain Voyager	Brain Innovation	Linux, macOS, Windows

Adapted from Poldrack et al [29]. *Using a "Virtual Linux Machine - VM" or the Windows Subsystem for Linux.

1.3.4.1. Pre-processing

The acquisition of fMRI data is not straightforward and a number of processing steps must be performed before the statistical analysis of the data. During image acquisition, movement can produce displacements of the data in time or space that must be corrected. Furthermore, artifacts of geometric distortion (mainly at air-bone interfaces) and signal drop out can lead to misregistration and loss of signal. The preprocessing steps aim to eliminate or reduce these artifacts and noise. Preprocessing is one of the most important steps in the pipeline of data analysis, as it will influence all the posterior statistical analysis. A small drawback can lead to error propagation, invalidating the interpretation of the results. The main standard processes, as stated by Poldrack [29], are schematized in Figure 16.

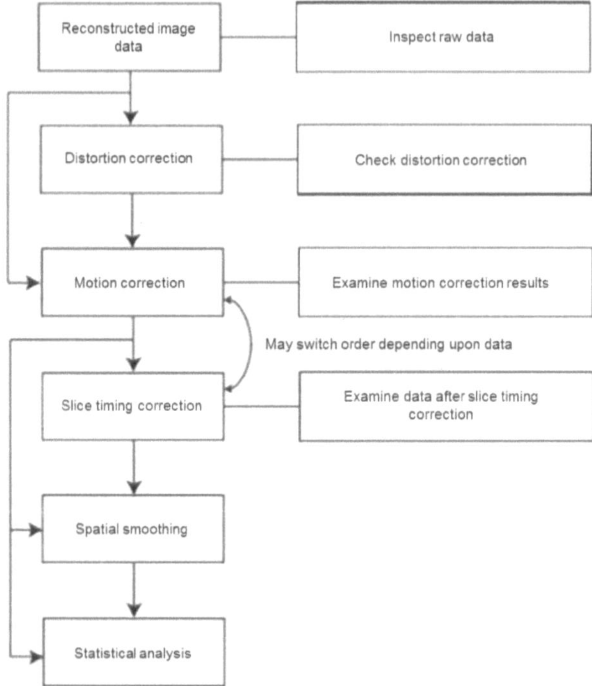

Figure 16: Common fMRI preprocessing steps including correspondent quality control checkpoints (on the right). (Adapted from Poldrack et al [29])

Distortion correction

As we have seen before, the BOLD fMRI data are acquired with echo-planar imaging technique, which is very susceptible to distortion artifacts due to static magnetic field inhomogeneities, especially in the air-tissue interface, like the sinuses and ear canals [57]. The distortion can assume two forms: dropout and geometric distortion [29].

Dropout corresponds to the reduced signal seen in brain regions adjacent to the air-tissue interfaces, like the fronto-orbital cortex (Figure 17) and lateral temporal lobe. Importantly, once the images are acquired, we cannot retrieve informative data from an area with significant dropout, so this question must be addressed during the acquisition in order to minimize the dropout. Along with the loss of the MRI signal on those regions, fMRI images can suffer geometric distortion in the same places.

Geometric distortion occurs as a result of errors in the location of structures on the final image because of the inhomogeneities of the magnetic field that will influence the spatial encoding. In this context, the geometric distortion appears along the phase encoding direction that is used, which is generally the yy (anterior-posterior) axis. The most commonly affected regions are the anterior prefrontal cortex and the fronto-orbital cortex.

Although we cannot recover the lost signal in dropout regions, we actually can attempt to undistort our images. It is possible to correct for the effects of magnetic field inhomogeneity using a field map, which characterizes the B_0 field [58]. MR phase is the most important quantity in a fieldmap sequence, whereas in normal imaging this phase is not of interest and is normally not saved when reconstructing the images. As a consequence, raw fieldmap scans are somewhat different from most scans and may contain images of complex values, or separate phase and magnitude images. Field heterogeneities can be suppressed by subtracting the phase of two images acquired at two different echo times. The difference in phase between the two images can be used to compute the local field inhomogeneity, and these values can then be used to create a map quantifying the distance that each voxel has been shifted [29]. The change in the MR phase from one image to the other is proportional to both the field inhomogeneity in that voxel and the echo time difference. The field value is therefore given by the difference in phase between these two images divided by the echo time difference.

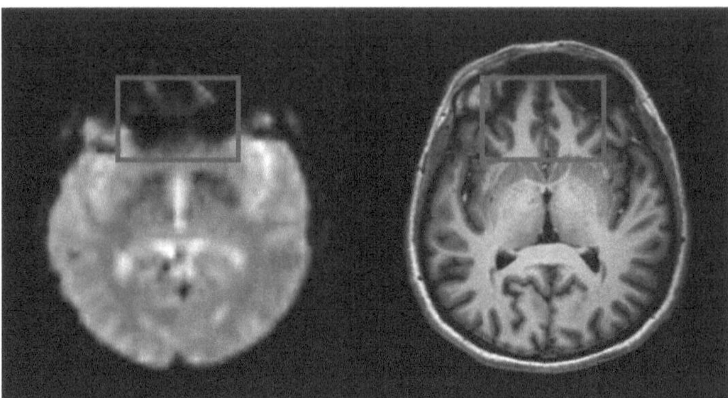

Figure 17: An example of signal dropout in gradient-echo EPI images. The EPI image on the left shows no signal on the fronto-orbital region (highlighted with the brown box) while we can see normal brain tissue in the same region on the T1-weighted image on the right.

The post-correction images with the field map must be visually inspected and compared to the pre-correction images to ensure that the distortion correction has not introduced new artifacts. Usually, the information on field map image can compensate for the geometric distortion artefacts, but not completely remove them. The artefacts are compensated by unwarping the EPI images and applying a cost-function masking in registrations to ignore areas of signal loss.

Motion correction

Head motion is one of the most important sources of data misregistration. Head motion is a huge concern in fMRI studies since even very subtle movements during acquisition can translate into a major source of error [32]. While the movement between adjacent slices can be subtle, the result of the misalignment of successive slices can translate into *bulk head motion* leading to incorrect anatomical positions between voxels of subsequent images [59]. There are several strategies to reduce head motion during the scan including foam pads around the head, bite bars, custom-designed cushions, thermo-plastic face masks, etc. Nevertheless, despite these efforts, the head usually moves a little during scanning making the motion correction always needed. To correct for this movement, there are some pre-processing steps that should be applied. This generally is done by realigning the image of the brain obtained at each point in time back to the first image acquired or to the middle volume. There are several correction methods, being the six-parameter motion correction method the most common. In this method, the head is considered a rigid body that can only move in six ways (three translation and three rotation movements). The movements are calculated at each point in time to minimize the image difference between the realigned brain and the brain in its original position [35]. It is important to notice that this method does not correct for all the effects of the movement since they realign the scans but cannot remove the movement induced signal artifacts. In order to overcome this problem independent component analysis (ICA) methods are used [29]. More recently, prospective (or adaptive/real-time) motion correction approaches have been used, allowing adequate correction of spin-history effects and intra-volume distortions [60].

Slice-timing correction

A single volume of BOLD fMRI data, acquired during one TR, consists of different echo-planar images acquired sequentially on time, meaning that, for a given volume data, the images will not be acquired at the exact same time, but with a delay of a few seconds. Consequently, a neuronal event that takes place at the same time in different slices will be displayed at a slightly different time on multiple slices within the brain. Slice-timing correction corrects for this staggering order of slice acquisition. If data acquisition is done with a short repetition time (TR), this pre-processing step might not be needed. In the case of rs-fMRI, slice-timing correction exhibits a negligible effect [61].

Spatial smoothing

Before statistical analysis, it is also common to do digital smoothing of BOLD fMRI data in space. This is achieved by averaging the intensity value of each voxel with the values of its neighbours. As a result, spatial smoothing attenuates high frequency fluctuations between adjacent voxels. BOLD fMRI data are typically composed of time-series information from many thousands of individual voxels, which means that statistical analysis of the data implies the application of a statistical test to each voxel being studied. By smoothing the data in space, one reduces the number of independent statistical tests that are being performed [35]. To smooth the image a low-pass filter is applied to the image which eliminates high-frequency signals, maintaining low-frequency information

[61]. Therefore, by smoothing the image, one improves the signal-to-noise ratio (SNR) at the expense of little loss in spatial resolution. Furthermore, by smoothing the image we can also overcome residual differences in anatomy between subjects that might otherwise render common areas of activation non-overlapping, providing better registration results [62]. On the other hand, smoothing too much will decrease statistical sensitivity for small focal areas of activation. Moreover, brain voxels can be averaged with non-brain tissue/background, resulting in inaccurate signal intensity. In the end, the amount of spatial smoothing to perform can be difficult to determine. The recommended amount of smoothing is the double of voxel dimensions [61].

For spatial smoothing, a convolution of the three-dimensional image with a three-dimensional Gaussian filter is performed. The filter is characterized by its full width at half maximum (FWHM), with a larger value of FWHM representing a greater data smoothing. The smoothing is given by:

$$FWHM = \sqrt{FWHM^2_{intrinsic} + FWHM^2_{applied}} \tag{15}$$

Spatial normalization

In order to study a population of individuals, we must guarantee that we are studying the same areas within the brain across subjects. This is achieved by computationally warping the anatomical structure of the brain of one subject to match a template brain within a standard defined space. The problem is that sometimes intersubject variability in anatomy cannot be overcome by warping brains to a standard space. While two subjects may have neural responses at the same true cytoarchitectonic location, the position of this site with respect to other landmarks in the brain may differ between subjects, leading to the spread of these locations when data are converted to a standard space [35].

1.3.4.2. Functional connectivity analysis in resting-state fMRI

When analyzing rs-fMRI data, one can extract information on the function of a specific brain region (functional mapping) or analyze the information on the functional connectivity between different brain regions [63].

After the pre-processing steps, the statistical method to analyze the fMRI data must be selected. Unlike task paradigms, which aim to study BOLD signal changes that reflect the level of engagement in response to a specific stimulus or task, rs-fMRI studies do not depend on a temporal response to a stimulus [32]. There are several analysis methods to withdraw information from a rs-fMRI acquisition. A summary of these methods is resumed in Figure 18.

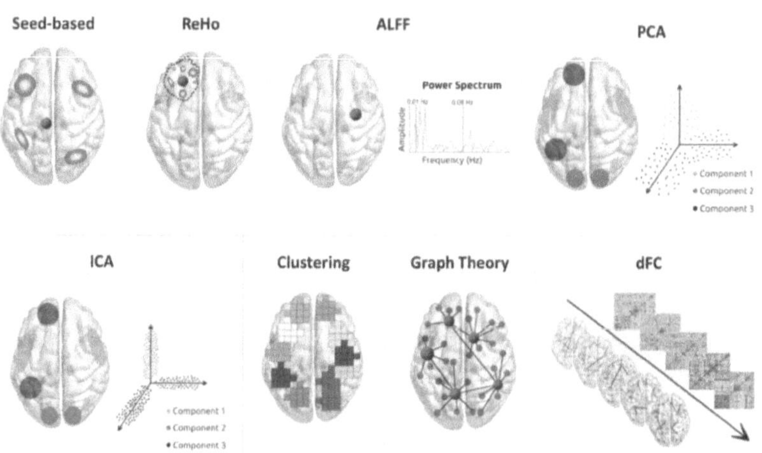

Figure 18: Adapted from Soares et al [61]. Resting-state fMRI methods. Top row (from left to right): seed based correlations; Regional Homogeneity (ReHo); Amplitude of Low Frequency Fluctuations; Principal Component Analysis (PCA). Bottom row (from left to right): Independent Component Analysis; Clustering; Graph Theory; dynamic Functional Connectivity (dFC).

Analytic approaches can be divided into two types: functional segregation and functional integration [64]. Functional segregation relates to the local function of a specific brain region and is, therefore, mainly used for brain mapping, reflecting the analysis of rs-fMRI activity. This type of analytic approach includes the ALLF and ReHo that we will discuss further. On the other hand, functional integration focuses on functional correlation or connectivity between different brain regions, so it assesses the brain as an integrated network, reflecting the analysis of rs-fMRI connectivity. This analytic approach includes the PCA, ICA, Graph analysis and seed-based connectivity analysis that will likewise be discussed further on.

These analysis methods can also be divided into two main groups: (i) Model-based and (ii) Data-driven methods, based on whether they use or not prior assumptions about the data, respectively [65]. Model-based methods require prior information, once fMRI data is compared to a predetermined model. On the other hand, data-driven models, also called model-free methods, do not rely on previous assumptions, since they are able to identify and extract the data. These latter models select the useful information and retrieve the data without constrictions [66].

For resting-state experiments the data-driven models are preferred since they do not require prior knowledge about the spatial and temporal activation patterns across the brain. In this way, they are also called exploratory methods. For validation, these methods are usually used together to analyze the data.

Model-based analysis

1) Seed-based correlational analysis

Historically, this was the first model used to study rs-fMRI [67]. This method is based on the selection of one seed (that can be a single voxel or a selected volume of the brain/region of interest - ROI), and comparison of its BOLD signal time-series to the time-series of all the other voxels in the brain [68]. Several metrics (eg, the cross-correlation coefficient, partial correlations, multiple regressions, and synchronization likelihood) can be used to assess associations between time-series of brain areas [63]. The activation of different brain areas at the same time suggests that these regions are functionally connected. The comparison can be made with (i) general linear model (GLM) or (ii) correlation analysis.

(i)   The GLM is an important tool for several fMRI analysis. It establishes a model and fits it to the data. The rationale behind GLM is finding the relation between a determinate response (dependent variable) and some predictors (independent variables). It consists in the modulation of the observed signal in terms of one or more regressors, also called explanatory variables (EVs). The general equation describing this model is shown below:

$$Y = \beta_0 + \beta_1 x_1 + \cdots + \beta_n x_n + \varepsilon \tag{16}$$

In this way, the fMRI signal (Y) can be explained by the sum of different components/variables ($x_1, x_2, \ldots, x_n$) weighted by a factor ($\beta_1, \beta_2, \ldots, \beta_n$) and a parameter of random noise (ε). The GLM can include variables that are nuisance regressors (components that interfere with the results but do not contain useful information), allowing to decrease the error associated with these regressors (usually head motion or age effect). Once the GLM is constructed, it is possible to mathematically compute the values for the weighting factors (or effect sizes) corresponding to each variable. The effect size associated with a given regressor quantifies how strongly the data is explained by that regressor. Hence, by introducing the BOLD time-series of the seed as a regressor in the GLM, it is possible to investigate which other regions of the brain are significantly related to the seed region. This regressor can be obtained by averaging the time-series of the all voxels within the seed region [69].

(ii)   The correlation analyses allow the computation of the degree of temporal synchrony between two different voxels. To assess this synchrony, a correlation coefficient (r) is calculated as:

$$r = \frac{\sum (x_i - \bar{X})(y_i - \bar{Y})}{\sqrt{\sum (x_i - \bar{X})^2 \sum (y_i - \bar{Y})^2}} \tag{17}$$

where r can range from -1 to 1, with $r = -1$ meaning perfect anti-correlation, $r = 1$ reflecting perfect correlation and $r = 0$ denoting no correlation. Naturally, this correlation analysis can only be done if only one hypothesis is being tested. In the resting-state studies, the bigger the

r (closer to $r = 1$), the stronger the association and thus the connectivity between two voxels (or two regions) [32].

Seed-based methods are easy to implement and the statistics behind this analysis are also simple to understand. Like mentioned before, and similar to other model-based analysis, the seed-based method implies the previous selection of an ROI, introducing some bias to the system. For rs-fMRI, ROI selection can be done using the data itself (each RSN can be extracted from a specific associated ROI) [70].

2) Regional Homogeneity analysis (ReHo)

Regional homogeneity analysis is a voxel-based method that compares one single voxel activation to the time-series activation of its neighbours. The statistics used to extract this correlation is the Kendall's coefficient of concordance that allows the measurement of the synchronization of the time-series of one voxel and the closest neighbours (based on a previously selected ROI) [71]. The ReHo method is easy to implement and interpret and is normally applied to rs-fMRI determinations [72].

3) Amplitude of Low-Frequency Fluctuations (ALFF)

The Amplitude of Low-Frequency Fluctuations and more recently the fractional ALFF (which is less sensitive to physiological noise) evaluates the signal amplitude or magnitude on a voxel by voxel basis. Brain slow (low-frequency) fluctuations are a distinctive pattern of activation that is very useful to study the resting-state networks. The magnitude of these fluctuations vary from subject to subject and can be markers of abnormalities [73].

Data-driven analysis

1) Principal Component Analysis (PCA)

The Principal component analysis relies on finding a set of orthogonal axes (called the principal components) that can potentially explain the variation on functional data, allowing the separation of the relevant information from noise [74, 75]. Three conditions are necessary to apply PCA: (i) a high SNR and (ii) linearity and (iii) orthogonality of the principal components. The directions that are included in the analysis are the ones that contain the majority of the variation [76].

2) Independent Component Analysis (ICA)

Independent component analysis is an extent of the PCA and is the most widely used method to process rs-fMRI data. This method separates each individual element in its underlying components, detecting the spatiotemporal structure of the signal by decomposing the BOLD signal into separate components. These components are then linearly mixed. The rs-fMRI data is modeled as a constant number of components that are spatially or temporally independent. ICA generates a set of spatial maps and corresponding time-courses [77]. For rs-fMRI analysis, ICA maps are usually spatial maps reflecting the spatially independent components. Temporal ICA contains fewer data points (when compared to the number of voxels of the spatial ICA), reason why the use of time-domain is deprecated. Spatial ICA allows better statistics estimations [78].

The spatial ICA algorithm is applied to the rs-fMRI dataset as depicted in Figure 19. The 4D rs-fMRI data (information within voxels over a period of time) is reorganized in a 2D matrix (voxels x time), where the voxels for a given time point are lined on a row, meaning that each single row corresponds to a 3D image [79]. This matrix is then re-arranged to obtain two new matrixes: one with the time-course of a specific signal/component in each row (components x time) and the other with a spatial map for each component (components x spatial maps). This last matrix gives rise to an image that we use to assess the connectivity.

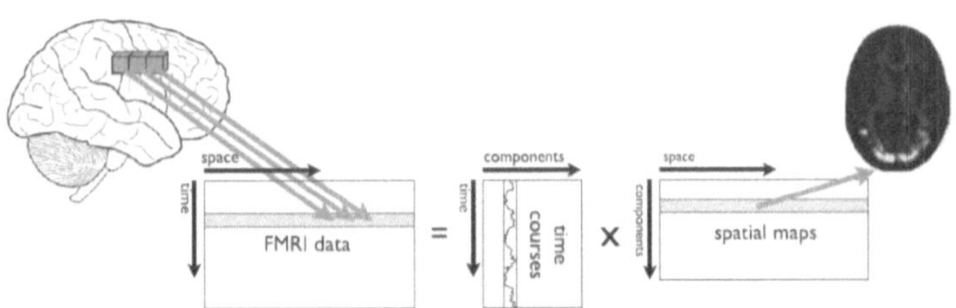

Figure 19: (Adapted from Beckmann et al [79]) Spatial ICA. Schematic illustration of the data representation and the spatial decomposition performed by spatial ICA on fMRI data.

Each functional network is reported as a spatial map of the z scores derived from the correlation between the time-series of each voxel and the mean time-series of that brain network [63]. The average z score of each network reflects the magnitude of functional connectivity within the network.

This technique is also sensitive to motion and physiologic signals (non-neural fluctuations), like cardiac, breathing and CSF-related pulsations. The identification of these components enhances the possibility to exclude them from the analysis, with ICA being a robust method to denoise the signal. However, the selection of components of interest is not trivial and is usually performed by visual inspection or correlation with previously defined RSNs templates. The user selection of the relevant components introduces great bias to the analysis [32]. On individual single-subject analysis (first-level analysis) ICA is relatively easy to implement, but complexity emerges when applying it to group analysis [80].

3) Clustering methods

The clustering methods are based on mathematical algorithms that group the data into clusters (subsets) [81]. These clusters include data which parameters are more similar to each other than compared to other clusters (for example, voxels can be grouped in the same cluster if they show temporal synchrony). This data-driven method allows the grouping of brain voxels with the same connectivity in the same cluster. As a single-voxel technique, in this method, every voxel in the data has the potential to be explored [82].

Clustering can be implemented using different mathematical algorithms: (i) hierarchical clustering, which defines an individual voxel as a cluster and combines similar clusters within a settled distance measuring [65]; (ii) partitional clustering, where the dataset is parted into non-overlapping clusters (a priori definition of the number of clusters to be set); (iii) spectral clustering, which uses graph representation (explained in detail next), assuming that voxels are nodes that are connected thought weighted edges, with the weights being similarity measures [83]. The major concern regarding cluster analysis is to assure the reproducibility of networks across subjects and to guarantee the uniformity of an individual network [84].

4) Graph theory

Graph theory is an increasingly promising technique for the study of brain connectivity and functional brain networks. This method interprets the brain as a network of nodes (voxels or regions, usually defined as ROI) linked by edges (the connections between the nodes, that can be, for example, time-series correlations) [85]. With this method is possible to study the functional connection between any regions within the brain, constituting an extension of the seed-based analysis with multiple (and in every desired region) seeds. The whole-brain network can be integrated as a mathematical model (assumed as a graph), and every topographic property of the brain network can be represented by graph theory metrics [86]. These metrics include (i) clustering coefficient - the number of clusters close to the selected node compared to all the possible connections; (ii) characteristic path length – the average of the shortest path between two nodes; (iii) centrality – number of the shortest paths that are linked to a certain node; (iv) modularity – measures the capability of parcellating a network into modules. The centrality of a node defines the extent of connections that it establishes with other nodes, and a high centrality means that the node is very informative within the network [86]. The more modular a network is, the tighter are the relationships within the module but looser are the connections between different modules [87].

For simplicity one can just describe the edges of the graph, relegating the characterization of the topological properties of the entire network.

5) Dynamic Functional Connectivity (dFC)

Dynamic functional connectivity evaluates the temporal components (signal fluctuations over time) of the spontaneous BOLD signal. dFC has the advantage of better identify the constant changes in patterns of neural activity and these changes in functional connectivity over time may reveal important information about brain networks [88]. This technique can be implemented with different approaches and metrics that will not be discussed here but can be found in Soares et al work [61]. The biggest disadvantage of this method is the complexity of statistical analysis.

6) Functional Connectivity Density (FCD) analysis

Functional connectivity density is the most basic measurement of functional connectivity as it identifies the highly connected functional hubs [89]. FCD reveals how strongly a voxel is connected but gives no information on the regions with which the voxel is connected. FCD analysis calculates the correlation of the BOLD time-series between each voxel and all the other voxels in the brain, allowing the calculation of FCD maps (with the cutoff distance being around 75mm) [90, 91]. FCD analysis is simple to apply and it does not need any model assumption. Although it can reveal the importance of functional hubs on brain connectivity it does not indicate which regions are connected.

2. MATERIALS AND METHODS

2.1. Participants

The participants of this study included 10 healthy volunteers from the community, 5 males and 5 females, with a mean age of 39.6 years old, ranging from 29 to 51 years-old. None of the participants had structural alterations on brain MRI and none of them was under medication. The study was approved by CHUP ethical committee and informed consent was obtained for all participants.

2.2. Imaging data acquisition

For this study, image acquisition was performed on an MRI scanner Achieva 3.0T TX Software release 5.4.1.1 (Phillips Medical Systems). All the images were acquired in the Neuroradiology Department at CHUP, during the development of the present book. The images were obtained under the supervision of the author and with the collaboration of the technical staff.

All data were acquired using a 32-channel phased-array head coil. In the scanner, foam cushions and earplugs were used to limit head motion and reduce scanner noise, respectively.

Both T1-weighted and T2-weighted structural images, fieldmap, and functional images were obtained for each patient.

T1-weighted images consisted in a 3D T1-TFE sequence with the following parameters: $TE = 2.948\,ms$, $TR = 6.557\,ms$ and flip angle of 8°. The images had a voxel resolution of $1 \times 1 \times 1\,mm^3$, square FOV of $240\,mm$ with a reconstructed matrix size of 512×512. The 3D T1-TFE sequence is a fast 3D gradient echo pulse.

T2-weighted images were based on a 3D FLAIR sequence with the following parameters: $TE = 340\,ms$, $TI = 1650\,ms$, $TR = 4800\,ms$ and flip angle of 90°. The acquisition voxel resolution of these images is $1.11 \times 1.11 \times 1.12\,mm^3$ (reconstructed to $0,88 \times 0,88 \times 0,56\,mm^3$) using a square FOV of $240\,mm$.

fMRI data were acquired using a gradient-echo EPI sequence (2D) with the following parameters: $TE = 33\,ms$; $TR = 2500\,ms$; flip angle of 88°. Sense factor of 2,5 and Bandwidth of $2743\,Hz/pixel$. Voxel resolution of $3 \times 3 \times 3\,mm^3$; matrix size of

$240 \times 80 \times 80$. Total scan time duration was $6.5\ minutes$. Each acquisition has 150 time points/slices. To guarantee steady-state magnetization, a total of 4 dummies were acquired and discarded. The participants were instructed to stay awake, not to think of anything in particular, to keep the eyes closed, and to stay still as much as possible, with shallow breaths.

An additional sequence, the field map, was collected to improve the co-register. Field Map is also a gradient-echo sequence, being a T2*-weighted image. For this data, Field map parameters were like following: $TE = 2,3/4,6\ ms$; $TR = 20\ ms$; flip angle of $10°$; voxel resolution of $3 \times 3 \times 3\ mm^3$; matrix size of $240 \times 80 \times 80$.

2.3. Imaging Analysis

All image analysis was performed with the FMRIB Software Library (FSL 6.0.4) (https://fsl.fmrib.ox.ac.uk/fsl/fslwiki). FSL is a comprehensive library that includes tools specialized in the analysis and processing of MRI, fMRI and DTI brain imaging data; it was developed in Oxford, UK by members of the Analysis Group [92].

The acquired images in the MRI scanner are exported from the visualizer in DICOM format, which is not compatible with FSL. To overcome this limitation, the DICOM images must be converted to NIfTI (FSL input image format). To do so, we used the *dcm2niigui* tool provided by the neuroimaging visualization program *MRIcron* (https://www.nitrc.org/projects/mricron). This tool not only converts images to the proper NIfTI format but also reorients them for proper visualization in FSL.

2.3.1. Pre-processing

After all the images were imported to FSL, each fMRI dataset underwent a pre-processing pathway using FSL's FEAT tool (https://fsl.fmrib.ox.ac.uk/fsl/fslwiki/FEAT/).

1. Brain Extraction

   The first step was to perform brain extraction using the Brain Extraction Tool (BET) from FSL (https://fsl.fmrib.ox.ac.uk/fsl/fslwiki/BET) [93]. This tool removes non-brain tissue like the scalp and bones, accurately excluding non-informative non-brain tissue. BET was applied to the 3D T1-weighted structural image (Figure 20), and also to the fMRI BOLD data.

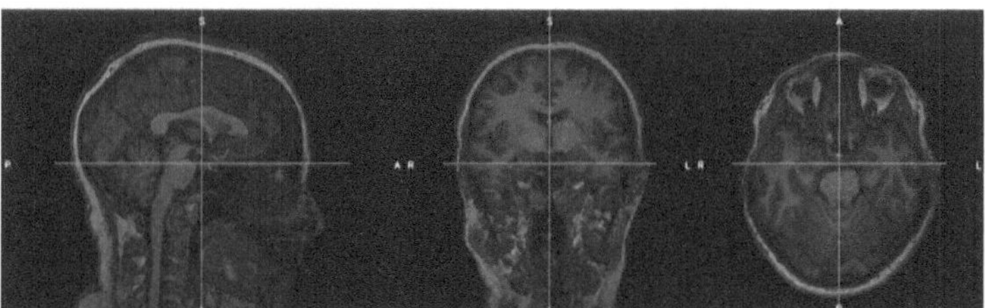

Figure 20: BET. T1-w structural image before (grey-scale) and after (red) brain extraction using FSL's BET. Sagittal, coronal and axial view. Images obtained in FSL's FSLeyes.

As we found the result from the default BET far from optimal in many cases, even with refined parameters, we included in the pipeline the script optiBET from Monti et al [94], with great improvement on extraction as we can confirm in Figure 21.

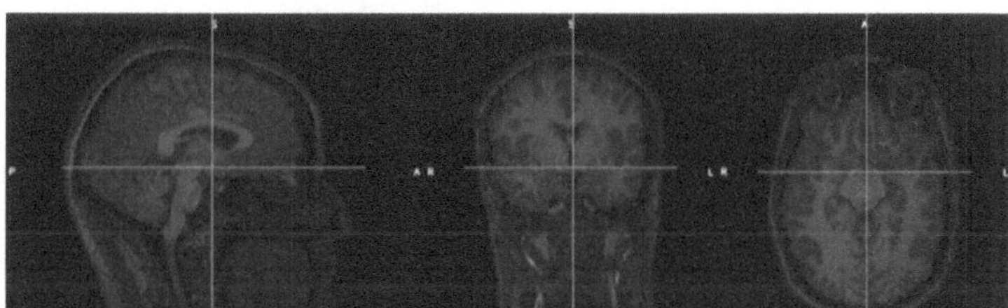

Figure 21: OptiBET. T1-w structural image before (grey-scale) and after (light red) brain extraction using the optimized script of optiBET for FSL's BET. Sagittal, coronal and axial view. Images obtained in FSL's FSLeyes.

2. Motion correction

Secondly, correction of head movements throughout the acquisition was accomplished by modeling displacement and changes in orientation of the images as rigid body transformations. Head motion artefacts were corrected with the tool MCFLIRT (https://fsl.fmrib.ox.ac.uk/fsl/fslwiki/MCFLIRT) [95], which implements an optimization method that uses FLIRT to apply rigid body transformations with 6 degrees of freedom (DOF) (3 rotations and 3 translations) to every volume, using the middle volume as an initial template. It computes the resulting mean displacement, as illustrated in Figure 22.

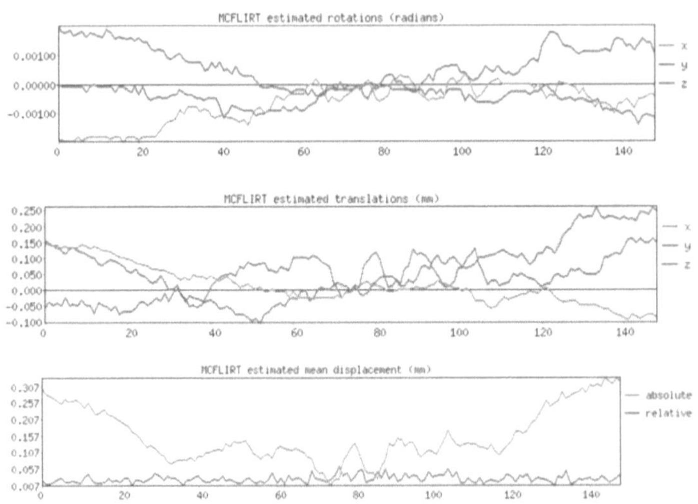

Figure 22: FEAT. Example of FEAT motion correction results, showing estimated rotations, translations, and mean displacement, in each volume, for one illustrative patient.

3. Spatial Smoothing

Spatial smoothing aimed to remove high-frequency fluctuations between adjacent pixels and was achieved applying a Gaussian kernel, with FWHM = 5 mm, individually in each volume of the fMRI data set. The usage of this low-pass filter increases SNR (signal-to-noise ratio) by removing noise without removing valid activation.

4. Temporal filtering

Temporal filtering was performed selecting a high-pass filter, to remove low-frequency drifts inherent to the acquisition process. These low-frequency artefacts were removed by using a local fit of a straight line, with $cutoff = 40TR = 100s$ which is the recommended value for rs-fMRI data with our acquisition parameters.

5. Distortion correction

To correct for distortion artefacts we applied fieldmaps. This is done with FUGUE tool (https://fsl.fmrib.ox.ac.uk/fsl/fslwiki/FUGUE) from the GUI which is part of the FEAT preprocessing options. FUGUE is a tool for EPI distortion correction in two steps: (i) geometrically unwarping the EPI images, and (ii) applying cost-function masking in registrations to ignore areas of signal loss.

Unfortunately, there is no standard sequence for fieldmap acquisitions and different scanners return different images. Normally, these images require processing before they represent images with field values in the desired units (of radians/second) in each voxel. The steps to construct the required fieldmap images from Phillips fielmap acquisition for B0 unwarping can be found on (https://osf.io/hks7x/) and are detailed in Appendix I. Different checkpoints for distortion correction with fieldmap and the final result of undistorted and distorted images are depicted in Figure 23 and 24, respectively.

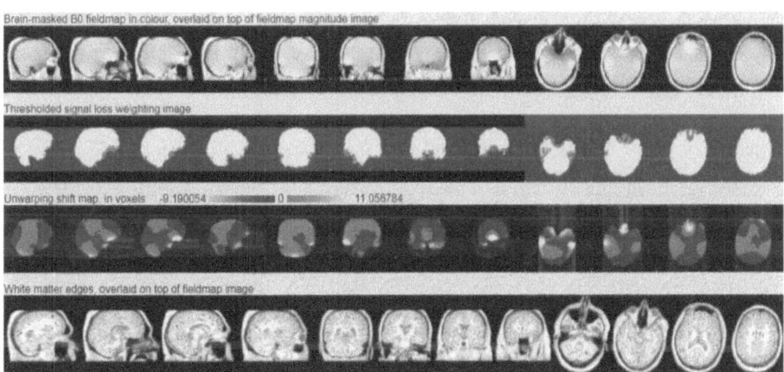

Figure 23: Sequential steps of distortion correction using fieldmaps. Top row: Brain-masked B0 fieldmap (orange), overlaid on top of fieldmap magnitude image; Second row: Thresholded signal loss weighting image; Third row: Unwarping shift map; Bottom row: White matter edges, overlaid on top of fieldmap image. Image obtained from unwarp report of FEAT data processing.

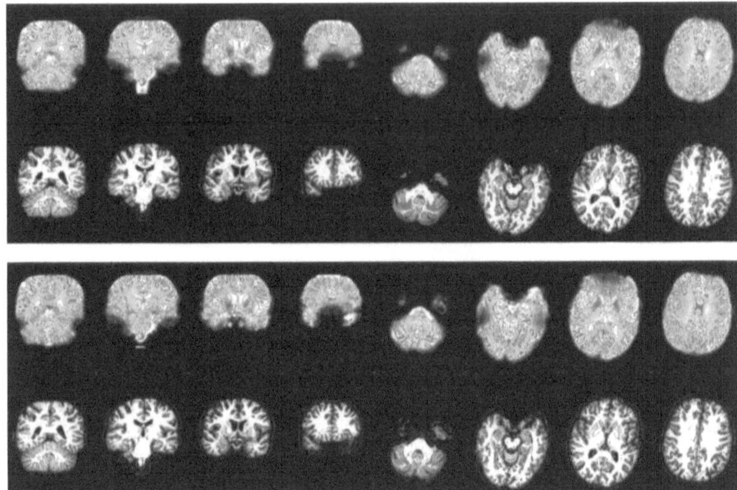

Figure 24: Registration without (first set of images) and with (second set of images) fieldmap correction. Top row of each image: co-register of white matter boundaries (red) to the example_func (fMRI); bottom row: co-register of white matter boundaries (red) to highres (structural T1).

When using the fieldmap information to correct for geometric distortion we need to select the phase encoding direction which, for our data, was the positive one. As we do not know this information *a priori* we should run both positive and negative phase-encoding directions and compare these results by visual inspection.

2.3.2. Registration

The pre-processed fMRI data were then co-registered with T1 structural images of each subject (Figure 15) using the FSL FLIRT (FMRIB's Linear Image Registration Tool) and FNIRT (FMRIB's Non-linear Image Registration Tool), FSL's tools for linear and non-linear registration, respectively.

The registration of fMRI images to the T1 acquisition was done using the boundary-based registration (BBR) method of FLIRT (https://fsl.fmrib.ox.ac.uk/fsl/fslwiki/FLIRT_BBR). BBR can detect the white matter (WM) boundary on EPI BOLD fMRI images, correlating it with the WM boundary of T1 and registering the images together [96]. After this step, the T1 structural image was then normalized to a standard MNI space (MNI152, standard-space T1-weighted average structural template image, $2 \times 2 \times 2\ mm^3$) (Figure 25). Normally, a simple linear regression is used to align the images (using the FLIRT tool from FSL - (https://fsl.fmrib.ox.ac.uk/fsl/fslwiki/FLIRT), but this strategy may not be fully accurate to align internal structures, since the brain of different subjects has different morphologies resulting from ageing, pathologies or race. To overcome this issue we can use a non-linear registration with FSL tool FNIRT (https://fsl.fmrib.ox.ac.uk/fsl/fslwiki/FNIRT/) after optimization of FLIRT. This approach improves the results, with better registrations.

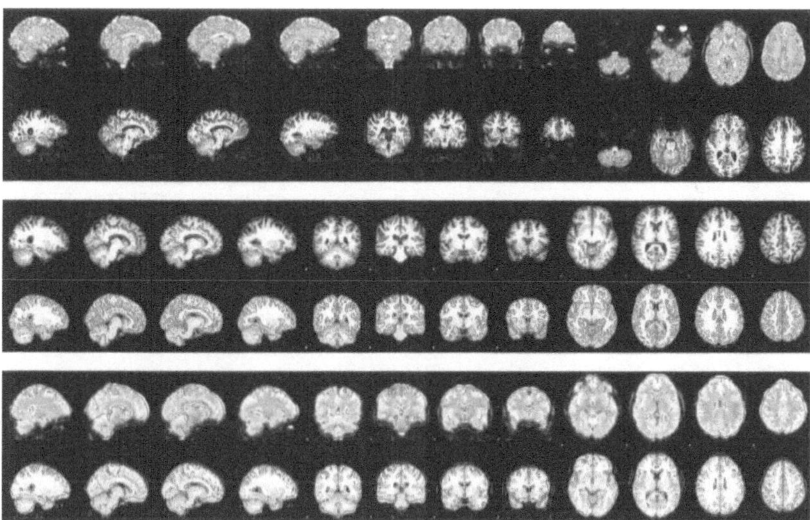

Figure 25: Example of registration of fMRI data to the structural image and the standard MNI space for one illustrative subject. Top image: functional space (fMRI) to high resolution structural space: (i) top line – fMRI image in grey and the high resolution structural image exhibited in red lines;(ii) bottom line – fMRI image depicted in red lines and the structural image in grey. Middle image: high resolution structural image to standard space: (i) top line – the structural image is in grey and the red lines represent the MNI; (ii) bottom line – structural image in red contours and the MNI in grey. Bottom image: combined transformation from functional to MNI space: (i) top line – fMRI is in grey and the red contours are from the MNI image; (ii) bottom line – fMRI is exhibited in red lines whereas the MNI is in grey.

2.3.3. Functional Connectivity Analysis

2.3.3.1. Independent Component analysis

The independent component analysis (ICA) was performed using Multivariate Exploratory Linear Optimized Decomposition into Independent Components (MELODIC) FSL tool (https://fsl.fmrib.ox.ac.uk/fsl/fslwiki/MELODIC/) to decompose the single-subject 4D datasets into sets of spatial components. MELODIC estimates automatically from the data the optimal number of components, which differ from subject to subject. The MELODIC offers three types of analysis: Single-session ICA, Multi-session temporal concatenation, and Multi-session Tensor-ICA. For the design of this work Single-session ICA (Figure 26) is the most suitable. This will perform standard 2D ICA on each of the input files (the 4D image of each subject being studied). The input data will each be represented as a 2D time x space matrix. MELODIC then decomposes each matrix separately into pairs of time courses and spatial maps [97]. The original data is assumed to be the sum of outer products of time courses and spatial maps. All the different time courses (one per component) will be saved in the mixing matrix melodic_mix and all the spatial maps (one per component) will be saved in the 4D file melodic_IC.

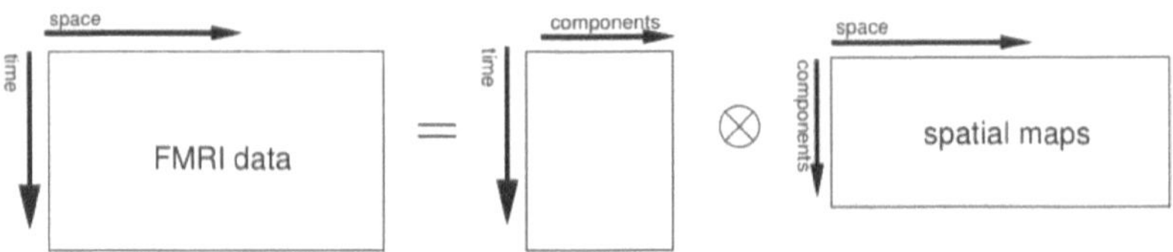

Figure 26: Schematic representation of MELODIC Single-session ICA.

Independent components (IC) classification

The Independent Components (IC) returned by MELODIC were classified, whether as noise components or neural signal components. This classification allows for further denoising of ICA, since the IC classified as noise can be removed from the input for further analysis, improving the results. For the classification, we used FSL's tool FIX (FMRIB's ICA-based Xnoiseifier) (https://fsl.fmrib.ox.ac.uk/fsl/fslwiki/FIX) [98] which is an automated approach for ICA-based denoising. In this way, the components of no interest can be removed from the reconstruction of the data, originating a denoised dataset. The general cleaning procedure in FIX consists of several steps: spatial ICA, component-wise feature extraction, classifier training, components' classification (predicting components' likelihood of being signal vs. noise, in new data) and denoising (removal of the artefactual components). The trained weight-file used was Standard.Rdata (supplied by FSL), which is indicated for fMRI studies with acquisition parameters close to the ones of this study.

FIX results were validated by hand, confirming that the noise and signal IC returned by FIX contained the expected characteristics [23]:

> • Spatial Maps: noise components' spatial maps hold a predominance of overlap brain boundaries, vessels, CSF and WM, whereas signal components contain well-defined areas;
> • Time-series: noise component's time-series show sudden jumps or sudden changes of oscillation patterns;
> • Power Spectrum: noise component's power spectrum contains high-frequency components, whereas signal components contain predominantly typical fMRI frequencies 0.01-0.08 Hz.

The IC classification can also be exclusively manually, by classification of each IC by visual inspection. To compare both methods we also performed manually classification to check for differences between both. The IC were evaluated by two neuroradiologists (blind to FIX classification) that rated them as being a true signal, a cardiac/CSF flow signal, and movement. When in doubt the components were classified as unknown in order not to be retrieved from further analysis. Manually classification was done on FSLeyes (https://fsl.fmrib.ox.ac.uk/fsl/fslwiki/FSLeyes) after the first MELODIC analysis (Figure 27).

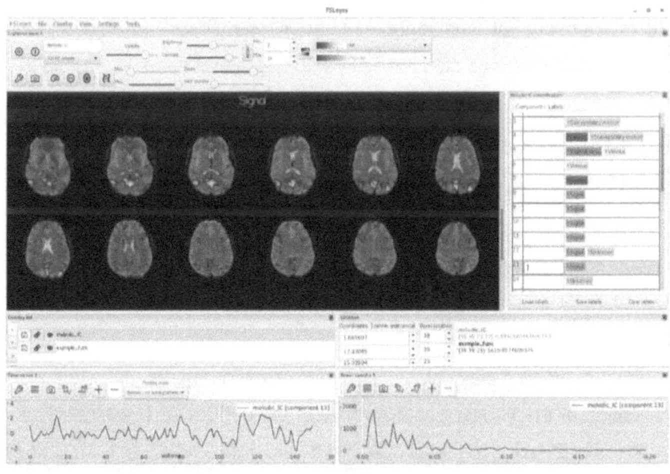

Figure 27: Manual classification of IC for one illustrative subject in FSLeyes.

After automated classification, the noise components returned by FIX were regressed out of the data to clean the activation maps, and all further processing was done using only the clean signal components from FIX.

RSNs identification

The cleaned signal was then used to run MELODIC again. The retrieved ICs from this analysis were the ones considered to further comparison with the RSNs described in the literature.

RSNs were first estimated by comparing the signal of the IC to the RSNs identified by *Smith et al* [46]. RSNs maps from *Smith et al.* were retrieved from BrainMap (https://www.fmrib.ox.ac.uk/datasets/brainmap+rsns/), and the IC of this study were identified using a template-matching procedure similar to that used in previous studies [99], in which IC are identified through spatial comparison against the RSNs templates. This is done by computing the Dice coefficient (DC) for all IC-RSN pairs [100]:

$$DC = \frac{2V_{overlap}}{V_{IC_1} + V_{IC_2}} \tag{18}$$

The DC provides a measure of similarity between the IC and the RSN template, varying between 0 and 1, higher values representing higher similarities. A DC over 0.3 is said to represent a good similarity between the IC and the RSN template [99]. Hence, each of the 10 RSNs in *Smith et al.* was attributed to the IC with which it shared higher similarity (higher DC), but only if it was also the RSN with which that same IC shared the highest similarity.

All RSNs were also assessed by visual inspection by using the matched maps published by *Smith et al.*

2.3.3.2. Seed-based analysis

RSNs identification

The Seed-based analysis was performed in this study in order to validate this method in the identification of the RSNs. For this analysis we performed two strategies: 1) using the FSL tools, and 2) using the Intellispace software from Phillips. In both cases we used the same 4 seeds (or regions of interest, ROI), based on state-of-art literature, corresponding to the posterior cingulate cortex (PCC), occipital pole (OP), middle frontal gyrus (MFG), and supplementary motor area (SMA). These seeds represent brain regions that are usually involved in important RSNs: the DMN, the visual, the frontoparietal network, and the sensorimotor, respectively. All the seeds used are well-defined areas within the respective network. The left and right frontoparietal networks were grouped into one individual network, the frontoparietal network [101].

1) FSL

For seed-based RSNs identification using FSL, 4 independent seed-based analyses were performed, each for the identification of one of the RSNs in *Smith et al.* A GLM analysis with multiple seeds as different regressors in a single design matrix was also performed.

In each analysis, a mask of the seed was created in the FSL's program FSLeyes (using the Harvard-Oxford Cortical Structural Atlases) and then binarized, by thresholding the image at 50%. After converting the mask from standard to functional space, the average time-series for the seed was extracted from the functional data and used as a regressor in the GLM. The multiple regression analysis was performed using the FSL's tool FEAT, and the Z statistic images returned were thresholded at Z=3 and compared with the RSNs in *Smith et al.* (also thresholded at Z=3). Importantly, high Z-scores in the Z statistic images reflect high effect sizes, and so are assumed to belong to brain areas with a strong correlation to the seed area and therefore are assumed to belong to the corresponding RSN.

2) IntelliSpace

The IntelliSpace Portal is an advanced visualization platform that allows the post-processing of several MR studies. Among the applications of this interface, is the MR iViewBOLD, an off-line package that facilitates the processing and interpretation for both block, event-related, and seed-based resting-state analysis. This tool presents the big advantage of automated pre-processing and registration to anatomical reference, enabling efficient workflow. Nevertheless, the system is restricted and closed, so few changes to the pre-processing of image analysis is permitted. The construction of the RSNs is based on SPM analysis. The program allows the selection of the seeds/ROI, which can be done with co-registration with the anatomic structural images. The selection and placement of the ROI are manual, and for each ROI selected, one pattern of activation is constructed and later displayed on the viewer. For each seed, a spherical ROI (5mm of radius) was manually designed. It is possible to change the threshold of smoothering.

Although very simple and intuitive to use, this tool can be quite limited in the application of rs-fMRI.

3. RESULTS

This study aimed to implement the evaluation of functional connectivity through resting-state fMRI in the Neuroradiology Department of Centro Hospitalar Universitário do Porto. Therefore, our results include not only the individual analysis of the constructed RSNs and the study of the functional connectivity for each participant but also the creation of a pipeline for performing rs-fMRI in our institution. This pipeline will serve as a guide and protocol for future clinical studies with patients.

In this chapter, we first describe our results relating to the exploratory analysis performed using ICA, as well as the results from the seed-based analysis, also including the network mean FC analyses. At the end of the chapter, we present a pipeline suggestion for implementation of rs-fMRI analysis.

3.1. Independent Component Analysis

FIX Denoising

FIX classified IC as signal components and as noise components for each patient and examples of the classification are demonstrated in Figure 28. The classification of all IC by FIX was visually inspected for validation (like described on methodology). Some of the IC classified was unknown by FIX were then reviewed by the neuroradiologists and further classified as shown in Table 3. The number of removed signals for each subject and the final number of the IC after the FIX clean-up is also characterized in Table 3.

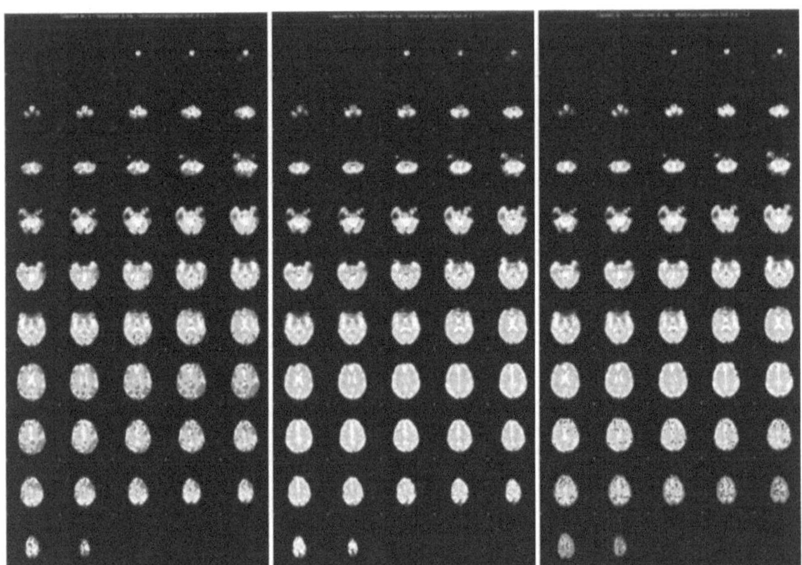

Figure 28: Examples of IC classified as signal (left), cardiac/CSF (middle) and movement (right) for one illustrative subject.

Table 3: Independent components before and after FIX and manual classification.

	Number of IC identified on first MELODIC analysis	Number (and identification) of IC classified as noise by FIX	Number (and identification) of FIX unknown IC classified as noise by hand	Final number of IC removed from the analysis
Subject 1	36	13 [1, 2, 8, 12, 13, 14, 15, 17, 24, 26, 27, 30, 31]	4 [6, 11, 20, 34]	17
Subject 2	36	9 [1, 3, 4, 5, 7, 8, 23, 33, 34]	2 [10, 12]	11
Subject 3	38	10 [1, 4, 5, 6, 7, 10, 15, 19, 20, 33]	0	10
Subject 4	37	6 [2, 4, 13, 15, 35, 36]	8 [1, 2, 3, 5, 6, 9, 24, 28]	14

Subject 5	25	14 [1, 2, 3, 4, 5, 6, 11, 13, 15, 17, 18, 19, 20, 24]	1 [7]	15
Subject 6	32	11 [2, 4, 5, 6, 7, 11, 13, 15, 16, 22, 24]	1 [3]	12
Subject 7	45	24 [1, 3, 4, 6, 7, 8, 12, 15, 17, 19, 24, 27, 28, 31, 32, 34, 36, 37, 38, 39, 41, 43, 44, 45]	3 [2, 5, 42]	27
Subject 8	35	16 [2, 3, 5, 6, 13, 14, 15, 19, 20, 22, 23, 24, 26, 27, 30, 35]	3 [4, 28, 31]	19
Subject 9	40	13 [2, 3, 4, 5, 6, 10, 16, 17, 19, 26, 28, 31, 39]	2 [1, 13]	15
Subject 10	34	7 [3, 6, 7, 12, 13, 14, 22]	5 [1, 4, 8, 11, 26]	12

As we have previously described, the FIX application needs a training data set. For better results, we can train our own data to construct a training data set with the specific characteristics of our MR machine and acquisition parameters. For this, we must hand classify all ICs to later construct the training data set, which we did. The "Training set CHUP" was constructed and tested in our population, but as mentioned in the methodology section, our analysis was done with the Standard.Rdata training data set to guarantee robustness.

RSNs Identification

The correspondence between the 10 RSNs in *Smith et al.* and the signal ICs of this study, along with the associated Dice Coefficient, is represented for each subject in Tables 4 to 13.

Table 4: Correspondence between the 10 RSNs in *Smith et al.* and 10 signal IC of Subject 1. Dice Coefficient (DC) computed for each pair.

Subject 1		
RSN from Smith et al	IC #	Dice Coefficient
visual medial network	IC 5	0,301
visual occipital pole network	IC 15	0,479
visual lateral network	IC 23	0,195
default mode network	IC 13	0,371
cerebellum network	----	----
sensorimotor network	IC 8	0,163
auditory network	IC 7	0,205
executive control network	IC 19	0,312
right frontoparietal network	IC 3	0,418
left frontoparietal network	IC 1	0,356

Table 5: Correspondence between the 10 RSNs in *Smith et al.* and 10 signal IC of Subject 2. Dice Coefficient (DC) computed for each pair.

Subject 2		
RSN from Smith et al	IC #	Dice Coefficient
visual medial network	IC 1	0,629
visual occipital pole network	IC 4	0,565
visual lateral network	IC 2	0,389
default mode network	IC 11	0,419
cerebellum network	IC 21	0,433
sensorimotor network	IC 6	0,275
auditory network	IC 5	0,255
executive control network	IC 5	0,225
right frontoparietal network	IC 19	0,379
left frontoparietal network	IC 20	0,348

Table 6: Correspondence between the 10 RSNs in *Smith et al.* and 10 signal IC of Subject 3. Dice Coefficient (DC) computed for each pair.

Subject 3		
RSN from Smith et al	IC #	Dice Coefficient
visual medial network	IC 19	0,455
visual occipital pole network	IC 14	0,355
visual lateral network	IC 14	0,357
default mode network	IC 1	0,453
cerebellum network	IC 16	0,110
sensorimotor network	IC 22	0,288
auditory network	IC 6	0,370
executive control network	IC 10	0,223
right frontoparietal network	IC 4	0,317
left frontoparietal network	IC 3	0,466

Table 7: Correspondence between the 10 RSNs in *Smith et al.* and 10 signal IC of Subject 4. Dice Coefficient (DC) computed for each pair.

Subject 4		
RSN from Smith et al	IC #	Dice Coefficient
visual medial network	IC 8	0,437
visual occipital pole network	IC 28	0,514
visual lateral network	IC 28	0,457
default mode network	IC 5	0,374
cerebellum network	IC 7	0,150
sensorimotor network	IC 14	0,342
auditory network	IC 3	0,319
executive control network	IC 29	0,229
right frontoparietal network	IC 16	0,375
left frontoparietal network	IC 20	0,342

Table 8: Correspondence between the 10 RSNs in *Smith et al.* and 10 signal IC of Subject 5. Dice Coefficient (DC) computed for each pair.

Subject 5		
RSN from Smith et al	IC #	Dice Coefficient
visual medial network	IC 8	0,102
visual occipital pole network	IC 17	0,140
visual lateral network	IC 17	0,129
default mode network	IC 1	0,396
cerebellum network	----	----
sensorimotor network	IC 10	0,168
auditory network	IC 13	0,101
executive control network	IC 12	0,188
right frontoparietal network	IC 6	0,382
left frontoparietal network	IC 2	0,353

Table 9: Correspondence between the 10 RSNs in *Smith et al.* and 10 signal IC of Subject 6. Dice Coefficient (DC) computed for each pair.

Subject 6		
RSN from Smith et al	IC #	Dice Coefficient
visual medial network	IC 5	0,308
visual occipital pole network	IC 21	0,413
visual lateral network	IC 5	0,469
default mode network	IC 3	0,468
cerebellum network	IC 18	0,216
sensorimotor network	IC 16	0,256
auditory network	IC 10	0,143
executive control network	IC 4	0,289
right frontoparietal network	IC 9	0,385
left frontoparietal network	IC 13	0,405

Table 10: Correspondence between the 10 RSNs in *Smith et al.* and 10 signal IC of Subject 7. Dice Coefficient (DC) computed for each pair.

Subject 7		
RSN from Smith et al	IC #	Dice Coefficient
visual medial network	IC 15	0,416
visual occipital pole network	IC 12	0,527
visual lateral network	IC 15	0,486
default mode network	IC 16	0,314
cerebellum network	IC 21	0,234
sensorimotor network	IC 4	0,214
auditory network	IC 17	0,371
executive control network	IC 11	0,215
right frontoparietal network	IC 3	0,314
left frontoparietal network	IC 10	0,318

Table 11: Correspondence between the 10 RSNs in *Smith et al.* and 10 signal IC of Subject 8. Dice Coefficient (DC) computed for each pair.

Subject 8		
RSN from Smith et al	IC #	Dice Coefficient
visual medial network	IC 4	0,241
visual occipital pole network	IC 1	0,283
visual lateral network	IC 13	0,457
default mode network	IC 5	0,474
cerebellum network	IC 14	0,110
sensorimotor network	IC 9	0,250
auditory network	IC 8	0,472
executive control network	IC 17	0,188
right frontoparietal network	IC 11	0,367
left frontoparietal network	IC 7	0,441

Table 12: Correspondence between the 10 RSNs in *Smith et al.* and 10 signal IC of Subject 9. Dice Coefficient (DC) computed for each pair.

Subject 9		
RSN from Smith et al	IC #	Dice Coefficient
visual medial network	IC 2	0,345
visual occipital pole network	IC 15	0,427
visual lateral network	IC 2	0,438
default mode network	IC 5	0,458
cerebellum network	IC 22	0,291
sensorimotor network	IC 11	0,306
auditory network	IC 21	0,241
executive control network	IC 12	0,238
right frontoparietal network	IC 3	0,303
left frontoparietal network	IC 6	0,335

Table 13: Correspondence between the 10 RSNs in *Smith et al.* and 10 signal IC of Subject 10. Dice Coefficient (DC) computed for each pair.

Subject 10		
RSN from Smith et al	IC #	Dice Coefficient
visual medial network	IC 13	0,461
visual occipital pole network	IC 7	0,343
visual lateral network	IC 8	0,305
default mode network	IC 18	0,414
cerebellum network	IC 26	0,204
sensorimotor network	IC 4	0,344
auditory network	IC 22	0,273
executive control network	IC 21	0,319
right frontoparietal network	IC 12	0,377
left frontoparietal network	IC 15	0,352

Generally, there was always a constructed IC that could represent an RSN, being very similar to the *Smith et al* RSN template on visual inspection. This also reflects on the DC values found for each subject that were globally very good. An exception is made for subject 5, where the DC was low for almost all networks, with DC>0.3 just for default mode and frontoparietal networks. In general, DC values were high for visual networks (maximum DC = 0,629), the default mode network, the frontoparietal, and, in less degree, for the sensorimotor network. For the auditory and executive control networks, the DC computed was small, which indicates that these networks may be scattered across multiple ICs (notice that there are much more identified ICs in our study than the 10 RSNs in *Smith et al*, meaning that some networks may split in different ICs. The same goes for the cerebellum network that was by far the most inconsistent network, with low DC in all participants and not able to be recognized in two patients.

Each of the IC-RSN pairs was manually confirmed by visual comparison of the respective spatial maps. In Figures 29 to 38 are presented examples of spatial maps for the 10 most relevant IC-RSN pairs.

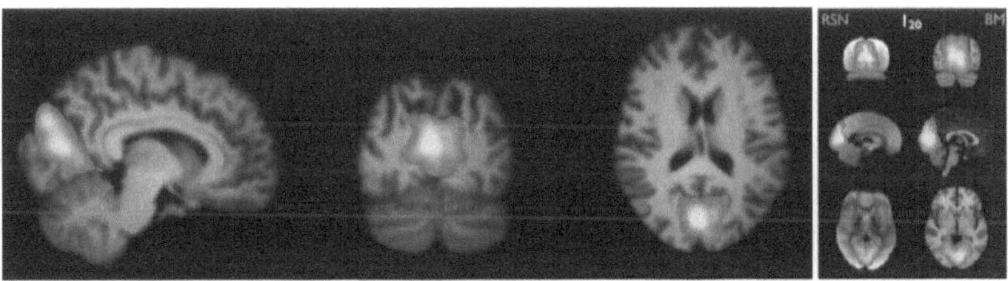

Figure 29: Spatial map of one IC representing the visual medial network (left) and the visual medial network identified in Smith et al (right). Spatial map was obtained in FSL's FSLeyes. All maps thresholded at Z=3.

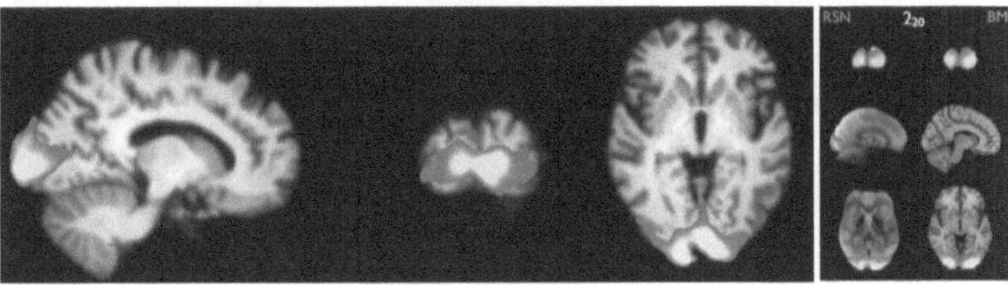

Figure 30: Spatial map of one IC representing the visual occipital network (left) and the visual occipital network identified in Smith et al (right). Spatial map was obtained in FSL's FSLeyes. All maps thresholded at Z=3.

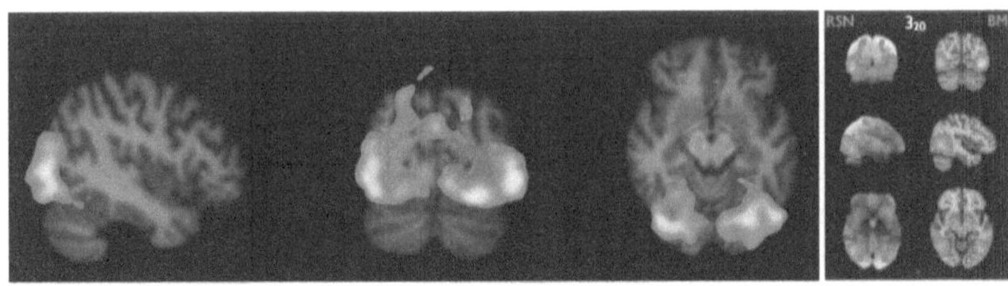

Figure 31: Spatial map of one IC representing the visual lateral network (left) and the visual lateral network identified in Smith et al (right). Spatial map was obtained in FSL's FSLeyes. All maps thresholded at Z=3.

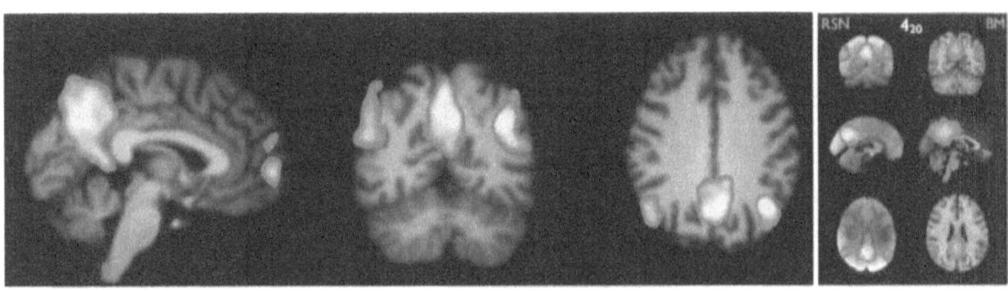

Figure 32: Spatial map of one IC representing the default mode network (left) and the default mode network identified in Smith et al (right). Spatial map was obtained in FSL's FSLeyes. All maps thresholded at Z=3.

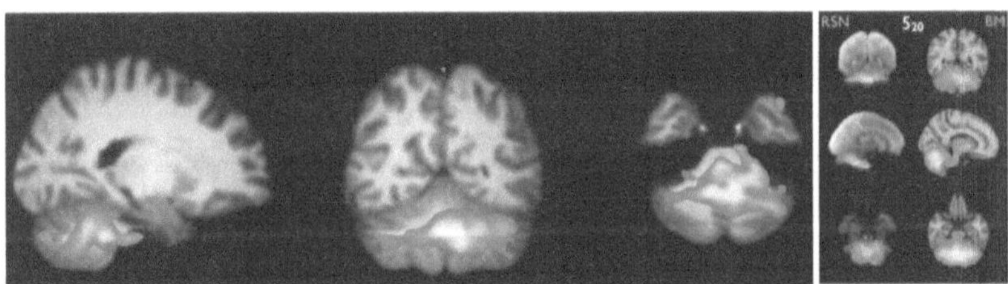

Figure 33: Spatial map of one IC representing the cerebellum network (left) and the cerebellum network identified in Smith et al (right). Spatial map was obtained in FSL's FSLeyes. All maps thresholded at Z=3.

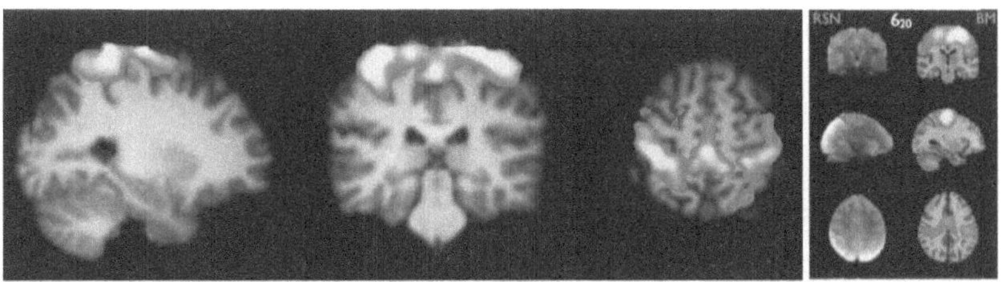

Figure 34: Spatial map of one IC representing the sensorimotor network (left) and the sensorimotor network identified in Smith et al (right). Spatial map was obtained in FSL's FSLeyes. All maps thresholded at Z=3.

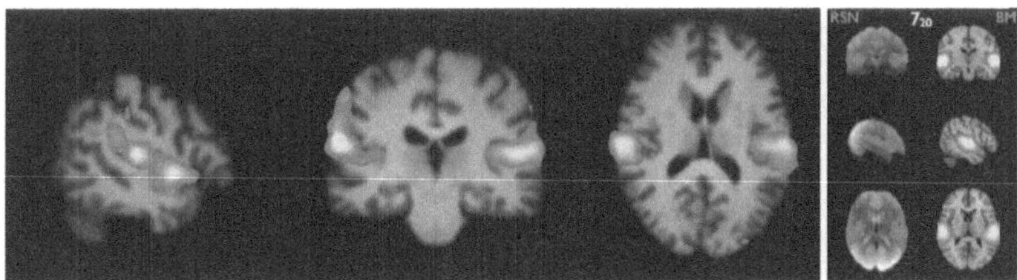

Figure 35: Spatial map of one IC representing the auditory network (left) and the auditory network identified in Smith et al (right). Spatial map was obtained in FSL's FSLeyes. All maps thresholded at Z=3.

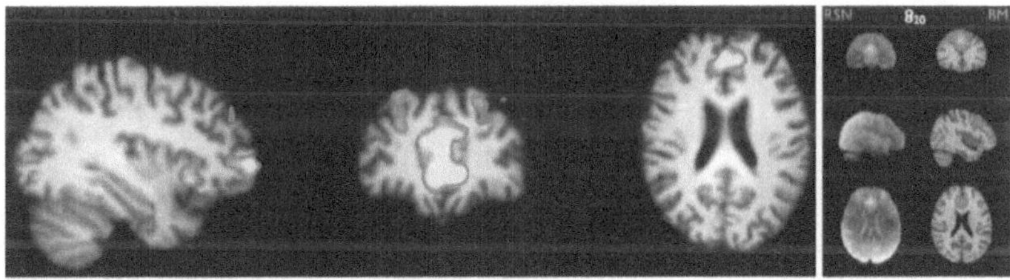

Figure 36: Spatial map of one IC representing the executive control network (left) and the executive control network identified in Smith et al (right). Spatial map was obtained in FSL's FSLeyes. All maps thresholded at Z=3.

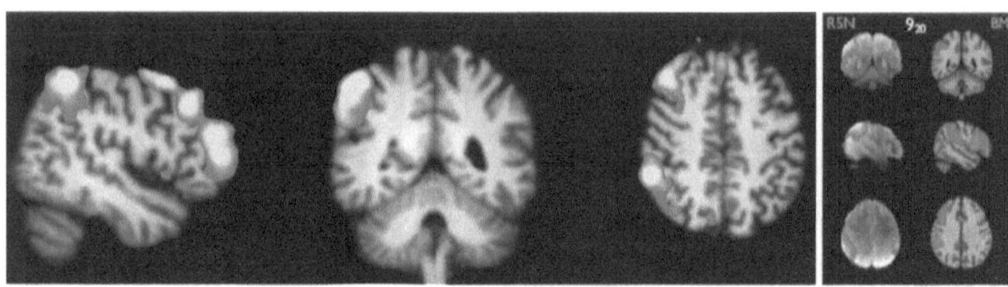

Figure 37: Spatial map of one IC representing the right frontoparietal network (left) and the right frontoparietal control network identified in Smith et al (right). Spatial map was obtained in FSL's FSLeyes. All maps thresholded at Z=3

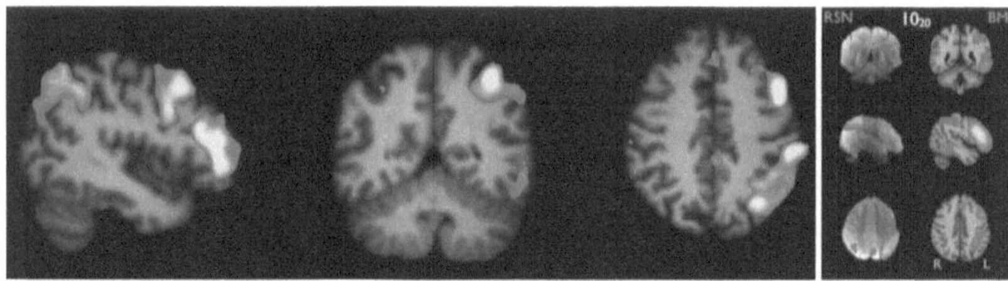

Figure 38: Spatial map of one IC representing the left frontoparietal network (left) and the left frontoparietal control network identified in Smith et al (right). Spatial map was obtained in FSL's FSLeyes. All maps thresholded at Z=3

3.2. Seed-based analysis

RSNs identification in FSL

For each subject, a multiple regression analysis was performed using the GLM framework. The mean BOLD signal time course of each seed was inserted as an explanatory variable within the design matrix and four contrasts of interest were defined, one for each EV. The output from this voxel-wise analysis was represented in Z statistic maps, one for each seed, reflecting the strength of the correlation between each voxel's time series and the seed's mean time course, for each subject. The spatial maps we obtained belong to the networks constructed based on the seeds considered to be the most relevant for this work as described in the methodology section.

We could construct all the four RSNs in all the patients, with varying strength among them. Both examples of GLM and resting-state networks constructed with seed-based analysis can be seen in Figure 39 and Figure 40-43, for one illustrative subject.

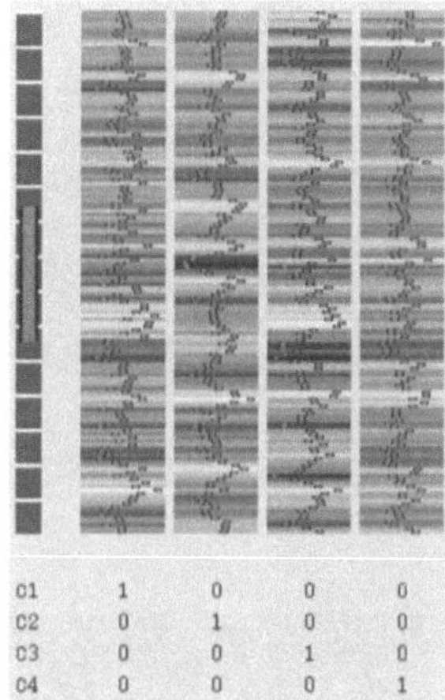

Figure 39: Design matrix used throughout the GLM analyses with the 4 explanatory variables (time course of each seed) and 4 t-test contrasts depicted.

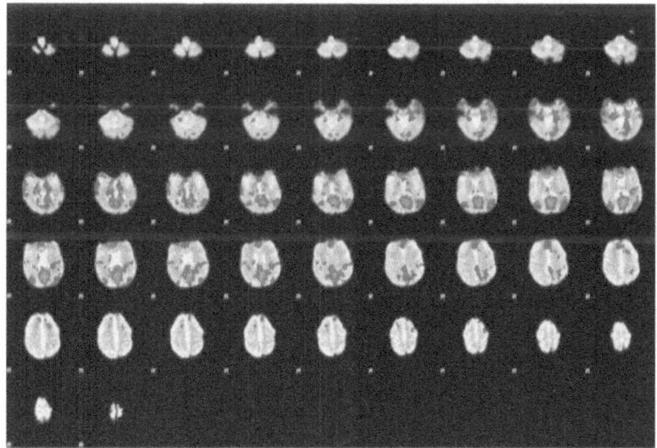

Figure 40: Spatial map of regions with high correlation to the seed PCC, obtained by seed-based GLM analysis with multiple regressors. Spatial maps was obtained in FSL's FSLeyes (thresholded at Z=3).

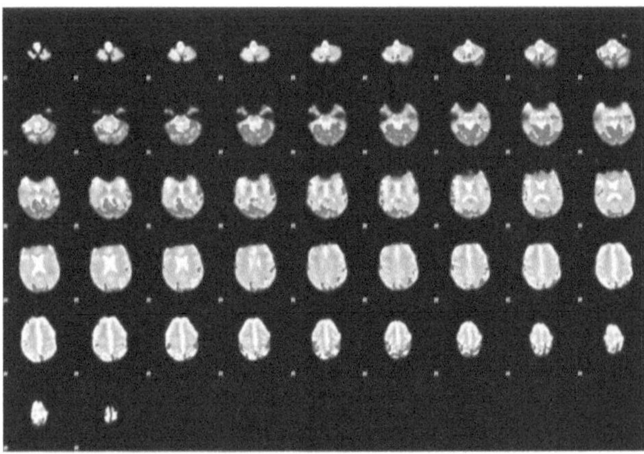

Figure 41: Spatial map of regions with high correlation to the seed OP, obtained by seed-based GLM analysis with multiple regressors. Spatial maps was obtained in FSL's FSLeyes (thresholded at Z=3).

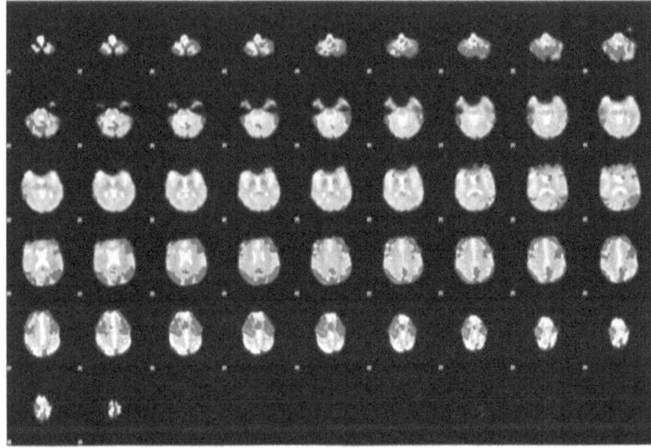

Figure 42: Spatial map of regions with high correlation to the seed MFG, obtained by seed-based GLM analysis with multiple regressors. Spatial maps was obtained in FSL's FSLeyes (thresholded at Z=3).

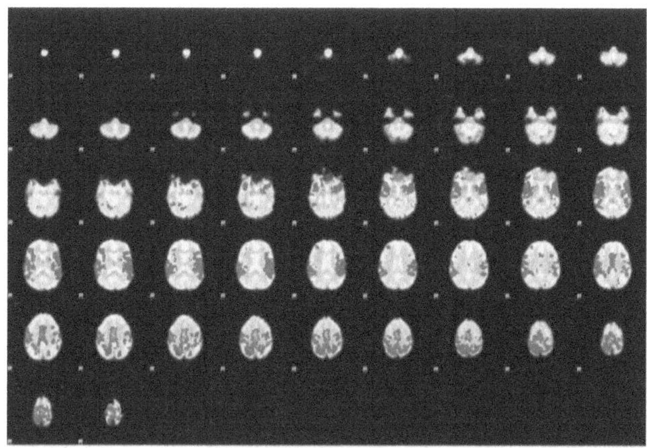

Figure 43: Spatial map of regions with high correlation to the seed SMA, obtained by seed-based GLM analysis with multiple regressors. Spatial maps was obtained in FSL's FSLeyes (thresholded at Z=3).

By visual inspection, the RSNs constructed based on the seeds satisfactorily corresponded to the RSNs templates described in the literature.

Mean connectivity

The seed-based analysis with the GLM approach also allows us to evaluate some metrics of brain connectivity. The mean functional connectivity (FC) was estimated in a single-level analysis as the mean Z-value from each Z statistic map for each selected seed (obtained as previously described). We retrieve this value for each participant and the results are in Table 14.

Table 14: Mean FC (measure from mean Z-value) from each seed for every participant.

	Posterior cingulate cortex	Occipital pole	Middle frontal gyrus	Suplementar motor area
Subject 1	4.756	3.746	3.032	3.455
Subject 2	3.650	4.438	3.248	4.023
Subject 3	4.563	3.295	4.279	3.452
Subject 4	4.588	3.614	3.126	3.842
Subject 5	3.514	3.843	3.842	3.197
Subject 6	5.382	3.121	4.512	4.281
Subject 7	4.296	4.192	3.387	4.296
Subject 8	3.729	3.121	3.103	3.191
Subject 9	4.756	3.822	3.098	3.126
Subject 10	5.018	4.395	3.256	3.338

Finally, the comparison of the mean FC of the different seeds is shown in Figure 44. The seed on the PCC was the one demonstrating the higher value of mean FC (mean Z-value), meaning that the DMN network was the network with the strongest correlation between the connected areas. The OP, MFG, and SMA showed lower and similar mean FC.

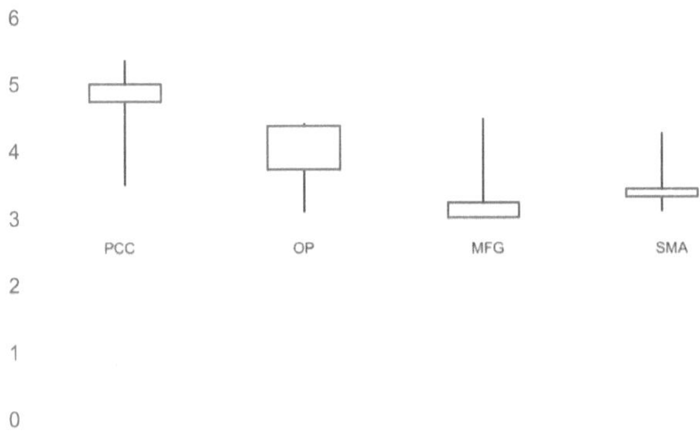

Figure 44: Blox-pot representing the distribution of the mean Z-values for each seed-based network.

RSNs identification in Intellispace

The RSNs identification with the seed-based package for rs-fMRI analysis of Intellispace was modest to say the best. We found the software very user-friendly, with easy and quick processing of the fMRI data, exempting all the pre-processing analysis (Figure 45). Nevertheless, the capability of the package to construct the RSNs was far from consistency and reproducibility, making this a fragile tool for clinical application (Figure 46).

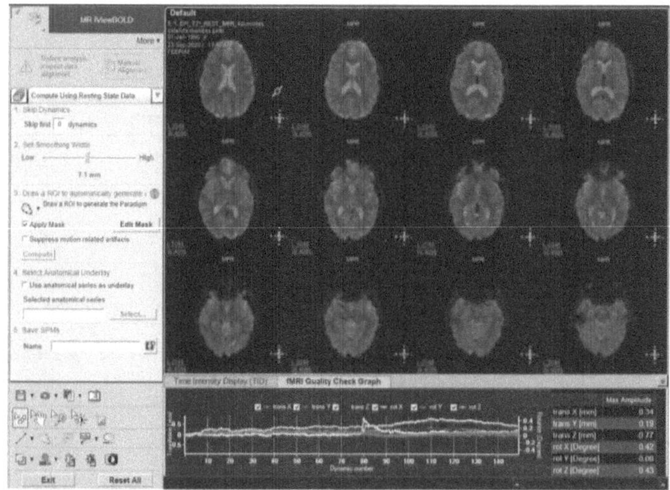

Figure 45: Display of the initial layout of MRI iViewBOLD of Intellispace. After uploading the image, all the preprocessing is automated. The software shows the Quality Graph Check but does not allow any modification on it.

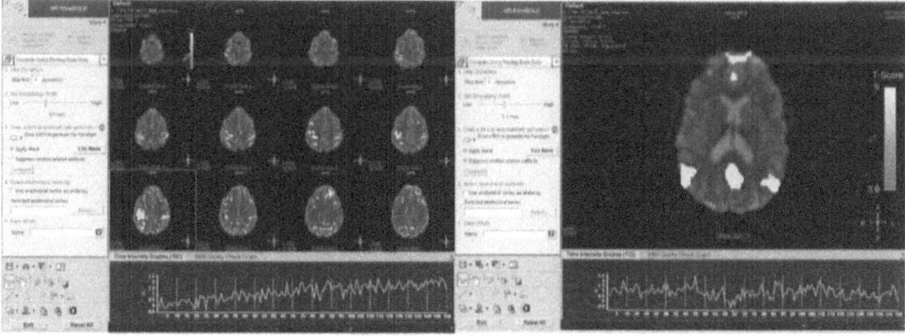

Figure 46: Examples of seed-based analysis with iViewBOLD: Motor activation on the right and DMN on the right.

3.3. Pipeline for Neuroradiology Department at CHUP

To perform a rs-fMRI analysis several steps must be taken into account. For simplicity and reproducibility, we have chosen the ICA-based analysis over the seed-based analysis to implement in our department (Figure 47). We now summarize the developed work in terms of the acquisition, pre and post-processing of rs-fMRI. A more detailed pipeline can be found in Appendix II.

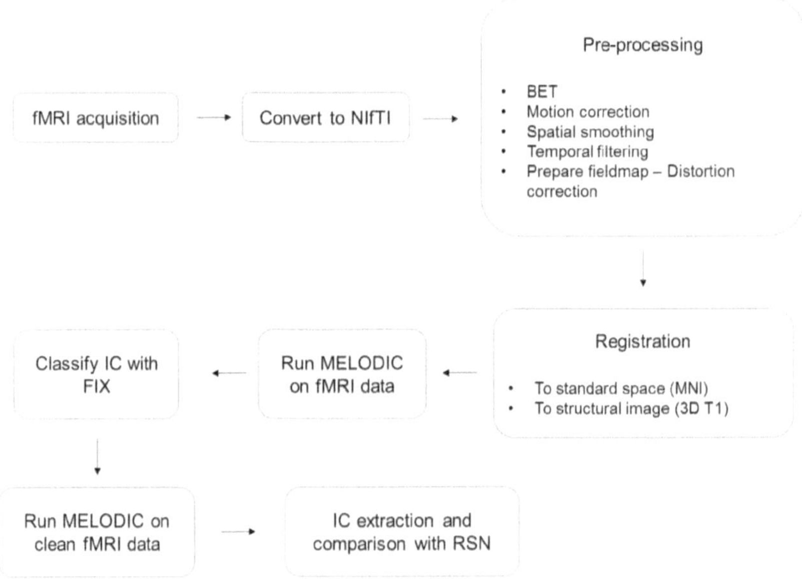

Figure 47: Schematic representation of the key steps to process rs-fMRI in CHUP.

4. DISCUSSION

All the evolution in the neuroscience field is brought alive with the cooperation of different professionals, with different but complementary assets and knowledge. The most difficult part of being a clinician eager to best understand the neurophysiology behind the neuroimaging is that there usually are no resources to develop advanced techniques.

4.1. Implementation of resting-state fMRI on healthy individuals in CHUP

The major ambition of this work was to implement a new neuroimaging technique to the Neuroradiology Department of CHUP. To achieve that goal we had to go a long way to attain the several steps that must be accomplished to have a thorough functional image working.

The first step was to develop the rs-fMRI acquisition for our Phillips machine, based on literature. We choose the parameters described in the methodology (Gradient-echo EPI sequence (2D): $TE = 33\,ms$; $TR = 2500\,ms$; voxel resolution of $3 \times 3 \times 3\,mm^3$; matrix size of $240 \times 80 \times 80\,mm$; flip angle of $88°$; duration: $6{,}5\,minutes$; 150 time points/slices), and we found our acquisition well-design, informative and reproducible.

This MRI sequence (with all the parameters defined) was stored in a common folder (rs-FMRI_LB_CMP) on the MRI protocols folder of our MRI scanner, being available for anyone wishing to use it. The recommendations to be given to the patients regarding the acquisition are the ones stated before.

We also constructed the fieldmap acquisition with the parameters described in the methodology section, but we found it of little added value for clinical use once it adds complexity to the pre-processing with little improvement on image quality and in the co-registration process. We analyzed the results regarding RSN identification with both the fieldmap corrected images and the images without distortion correction and we did not find any significant difference, meaning that, in our case, the suppression of the distortion correction with fieldmap images did not influence the results. As so, we believe that this step could probably be skipped if good care is taken in the EPI pulse sequence to optimize image quality and minimize its distortion. Structural images, 3D FLAIR and 3D

T1, are also on the same folder and should always be acquired to allow the identification of structural lesions and the registration steps, respectively.

The second step was to perform the image processing. All processing steps were done with FSL. FSL is directly supported in multiple hardware and software platforms like Linux, macOS and can also be deployed as a virtual machine in Microsoft Windows. It provides an advanced and readily available solution for rs-fMRI analysis, being license-free for non-commercial use. In the absence of available hardware and software in our department, we used a virtual machine from VMware Workstation Player. Because of administration restrictions on the computers installed in our department, there is only one workstation in which the virtual machine could be run. This will be, for now on, and until a new investment is done, the way to process the rs-fMRI acquisition in CHUP.

Along with FSL, some other tools needed to be installed, as is the case of FIX. FIX tool is a set of R, MATLAB and shell scripts, requiring various other software than just FSL, and for now, it is not bundled as part of FSL. We encountered some difficulties regarding FIX installation, but we finally manage to integrate it into our workflow. In addition to the standard training data set used, we also created our own training data set "Training set CHUP" with our participants. Although the application for this particular study is limited, the "Training set CHUP" will be useful in future studies, since the training data is more efficient the similar it is to the data being analyzed.

After going through all the technical details, we could achieve our goal of studying brain connectivity with rs-fMRI. All the careful pre- and post-processing of the data enable the congruent results. With the data-driven analysis, ICA, we were able to construct all the main RSNs described in the literature and we validated our results both with visual inspection (all spatial maps appeared sufficiently similar to the matched RSNs template) and with statistical analysis with Dice Coefficient. Regarding the seed-based GLM analysis, we also succeed in obtaining robust spatial maps for each selected seed that were further validated by visual comparison with the spatial maps of the same RSNs in *Smith et al.* That way, all the ROIs used for this study were confirmed to be reliable seed regions for the identification of the corresponding networks.

Our results regarding the mean FC showed that the Default mode network was the one with higher connectivity (through analysis of the Z maps of the PCC seed) when compared with the task-positive networks, like the visual, frontoparietal, and sensorimotor networks. These results are in line with the literature, where DMN is often described as the most important cortically based network during rest [102]. The study of the mean FC, although of little interest to compare the strength of the networks, is of major importance in further works aiming to compare the functional connectivity between groups of patients with some pathology or to compare disease groups to healthy controls.

The results indicate that both approaches under analysis (ICA and Seed-based) were able to identify RSNs with good enough efficacy for the purposes of this project. Thereby, we validated our results regarding functional mapping and brain connectivity with both the most used methods in the literature. The study of brain functional connectivity by rs-fMRI can now be performed in the Neuroradiology Department of CHUP, by simply following the proposed pipeline.

As the methodology for ICA is simpler in terms of post-processing (when compared with seed-based analysis), we propose that this should be the preferable approach for clinical practice application and for less experienced researchers.

4.2. Limitations

The main limitation of this work is the small sample size. Ten subjects are enough to bring some conclusions, but it is a limited representation of the population. Moreover, we recruit young healthy volunteers from the general population, most of them highly differentiated, which may not represent our future debilitated patients. We may face issues with the compliance of older patients, who usually have difficulties in laying still during the 6 minutes acquisition time.

Another concern is the effect of aging on mean functional connectivity and consequently on the construction of RSNs. It is now well established that, early in life, every primary sensory and cognitive network suffers some degree of age-related decline, including reduced within-network connectivity [103]. This can somehow influence the results if we are testing an older population, although the full extent of age modulation is not completely understood.

Lastly, it also constitutes a limitation to our work the fact that the image processing can only be performed in the virtual machine installed in the specific workstation. As an asset to the Neuroradiology department, ideally, we should have a dedicated set-up, with both hardware and software capable of more complex installations and pipelines. Only then, it will be possible to further improve and to translate the research expertise to the clinical knowledge.

4.3. Future Work

Along with the main purpose of this work, which was to implement and validate the rs-fMRI in healthy individuals in CHUP, the ultimate intention of this project was to give the basis and the background for future research works on the neuroscience field in our department. While working on a tertiary and university hospital we, as clinicians, have access to a great number of patients with a lot of pathologies, usually rare and differentiated, that present a big opportunity for learning. The step forward is to consciously use precious information gathered from rs-fMRI to improve our knowledge about the brain pathologies, how and why they develop, what is the expected course, how to understand the pathologic march and how to better predict outcomes or even the therapeutic response. From a daily clinician perspective, rs-fMRI is more important to give response to the individual question, in order to localize determinate brain functions that can be used, for example, to integrate the surgical planning. In this context, individual analysis, like the one we performed on the individual healthy subjects, is more valuable. On the other hand, when we think about big groups of patients with a specific disease or different subgroups of pathology that we want to study as a whole, the group analysis is more appropriate. Although the single-subject approach is the one that we will probably use the most in our daily clinical practice, this work leaves an open window to bigger and audacious group analysis that would allow comparison between groups of patients, may we have the support of our administration to engage investigational researches.

In future works, it would be interesting to compare the resting-state networks with task-related activity in order to understand if the networks extracted through functional connectivity by rs-fMRI tap into the complex activations subserving task execution. This would be essential to ascertain whether functional mapping with rs-fMRI is effective for pre-surgical planning.

On the image processing workflow, we can also improve our image analysis, namely in the pre-processing steps, including the application of additional motion regressors like the FSL Motion Outliers.

In the technical aspect, for future work, it would also be interesting to collect information on physiologic parameters (cardiac frequency, respiratory frequency) to better filter the noise components. The combination of the rs-fMRI results with other imaging modalities such as electroencephalogram could also be of great value, namely in differentiation neurovascular from neuroelectric phenomena of RSNs. In addition, the relation between functional (rs-fMRI) and anatomic (diffusion tensor imaging) connectivity can also be of interest to further understand normal brain connections and potential alterations in these relations in neurologic diseases [41].

5. CONCLUSION

This present book aimed to implement the brain functional connectivity analysis by rs-fMRI on the Neuroradiology Department of Centro Hospitalar Universitário do Porto. The objective was to integrate resting-state functional neuroimaging into the daily practice of neuroradiologists, as well as to create guidelines with all the necessary steps for the processing of rs-fMRI data using FSL, enabling its application by any neuroradiologist interested in doing so.

We could achieve our goal, and it is now possible to acquire and process rs-fMRI data in Paulo Mendo Neuroradiology Department, following the pipeline that we have proposed in this work.

We have demonstrated that our acquisition sequences were conceptually well designed, reproducible and robust. Also, we could implement all the pre- and post-processing steps, being able to actually construct the RSNs and calculate their connectivity in the studied individuals.

Although we did not provide any scientific advance to the neuroscience research community, the strength and value of our work resides in the fact that we could bring together professionals from very different backgrounds to work together and to implement a research and investigational tool to the everyday practice of a clinical hospital. This might seem of little value, but we truly believe that the answer for the development of the neuroradiology field is to cooperate with professionals from diverse backgrounds: statistics, computer science, engineering, psychology, mathematics, and physics. It is in the hospitals, and not in the labs, that are the people that make us want to go further – the ones in need to be helped, the patients. It is our mission to bring science closer to the patients and to join the powerful force of engineers and physicists with medicine.

6. REFERENCES

1. Gray H, Standring S, Ellis H, Berkovitz BKB. Gray's anatomy : the anatomical basis of clinical practice. 39th ed. Edinburgh ; New York: Elsevier Churchill Livingstone; 2005.

2. Ten Donkelaar HJ, Broman J, Neumann PE, Puelles L, Riva A, Tubbs RS, et al. Towards a Terminologia Neuroanatomica. Clinical anatomy (New York, NY). 2017;30(2):145-55. doi: 10.1002/ca.22809.

3. Arslan O. Neuroanatomical basis of clinical neurology. Second edition. ed. Boca Raton: CRC Press, Taylor & Francis Group; 2015.

4. Fix JD. High Yeld Neuroanatomy. 4th Edition. Philadelphia PA. Lippincott Williams & Wilkins; 2008.

5. Kiernan JA. Anatomy of the Temporal Lobe. Epilepsy Research and Treatment. 2012;2012:176157. doi: 10.1155/2012/176157.

6. Flores LP. Occipital lobe morphological anatomy: anatomical and surgical aspects. Arquivos de neuro-psiquiatria. 2002;60(3-a):566-71. doi: 10.1590/s0004-282x2002000400010.

7. Gogolla N. The insular cortex. Current Biology. 2017;27(12):R580-R6. doi: https://doi.org/10.1016/j.cub.2017.05.010.

8. McLachlan R. A Brief Review of the Anatomy and Physiology of the Limbic System. The Canadian journal of neurological sciences Le journal canadien des sciences neurologiques. 2009;36 Suppl 2:S84-7.

9. Lanciego JL, Luquin N, Obeso JA. Functional neuroanatomy of the basal ganglia. Cold Spring Harb Perspect Med. 2012;2(12):a009621-a. doi: 10.1101/cshperspect.a009621.

10. Koziol LF, Budding D, Andreasen N, D'Arrigo S, Bulgheroni S, Imamizu H, et al. Consensus paper: the cerebellum's role in movement and cognition. Cerebellum. 2014;13(1):151-77. doi: 10.1007/s12311-013-0511-x.

11. Edelman RR, Hesselink JR, Crues JV, Zlatkin MB. Clinical Magnetic Resonance Imaging. vol vol. 3. Saunders Elsevier; 2006.

12. Rinck P. Magnetic Resonance in Medicine. 4th edition: in chapter twenty: An Excursion into the History of MR Imaging, Wiley; 2001.

13. Morris GA. Spin dynamics: Basics of nuclear magnetic resonance, second edition. Malcolm H. Levitt. Wiley Chichester. 2008. pp xxv + 714. ISBN 978-0-470-51118-3(hbk) 978-0-470-51117-6 (pbk). 2009;22(2):240-1. doi: https://doi.org/10.1002/nbm.1356.

14. Slichter CP. Principles of magnetic resonance. 3rd enl. and updated ed. Springer series in solid-state sciences, vol 1. Berlin ; New York: Springer-Verlag; 1992.

15. Feynman RP, Basic Books (Firm), California Institute of Technology. The Feynman lectures on physics. v.19, Feynman on masers and light. v.20, Feynman on quantum mechanics and electromagnetism. New York Pasadena: Basic Books; California Institute of Technology; 2010.

16. Pooley RA. Fundamental Physics of MR Imaging. 2005;25(4):1087-99. doi: 10.1148/rg.254055027.

17. Liang Z-P, Lauterbur PC, IEEE Engineering in Medicine and Biology Society. Principles of magnetic resonance imaging : a signal processing perspective. IEEE Press series in biomedical engineering. Bellingham, Wash. New York: SPIE Optical Engineering Press;IEEE Press; 2000.

18. Edwards J. Principles of NMR. 2009 (Accessed on 08/2020) [ONLINE]. http://www.process-nmr.com/nmr1.htm.

19. Cowan BP. Nuclear magnetic resonance and relaxation. New York: Cambridge University Press; 1997.

20. Bushberg JT. The essential physics of medical imaging. 3rd ed. Philadelphia: Wolters Kluwer Health/Lippincott Williams & Wilkins; 2012.

21. Classical Response of a Single Nucleus to a Magnetic Field. Magnetic Resonance Imaging. 2014. p. 19-36.

22. Jp R Marques. Effects of dipolar fields in NMR and MRI [PhD book]. University of Nottingham; 2004.

23. Brown RW, Cheng Y-CN, Haacke EM, Thompson MR, Venkatesan R. Magnetic resonance imaging : physical principles and sequence design. Second edition. ed. Hoboken, New Jersey: John Wiley & Sons, Inc.; 2014.

24. Mazzola AA. Ressonância magnética: princípios de formação da imagem e aplicações em imagem funcional. Revista Brasileira de Física Médica. 2015;3(1):117-29. doi: 10.29384/rbfm.2009.v3.n1.p117-129.

25. Reimer P, Parizel PM, Stichnoth FA. Clinical MR Imaging: A Practical Approach. Springer; 2003.

26. Sleigh A. MRI basic principles and applications. 3rd edn, by Mark A. Brown and Richard C. Semelka. Wiley-Liss, Chichester, £32.50. 2004;17(4):209-. doi: https://doi.org/10.1002/nbm.864.

27. Lindquist M, Wager TD. 1 Principles of functional Magnetic Resonance Imaging. 2014. (Online textbook, available for download only).

28. Belliveau JW, Kennedy DN, McKinstry RC, Buchbinder BR, Weisskoff RM, Cohen MS, et al. Functional Mapping of the Human Visual Cortex by Magnetic Resonance Imaging. Science. 1991;254(5032):716-9.

29. Poldrack RA, Mumford JA, Nichols TE. Handbook of Functional MRI Data Analysis. Cambridge: Cambridge University Press; 2011.

30. Sirotin YB, Das A. Anticipatory haemodynamic signals in sensory cortex not predicted by local neuronal activity. Nature. 2009;457(7228):475-9. doi: 10.1038/nature07664.

31. Ogawa S, Lee TM, Kay AR, Tank DW. Brain magnetic resonance imaging with contrast dependent on blood oxygenation. Proceedings of the National Academy of Sciences of the United States of America. 1990;87(24):9868-72. doi: 10.1073/pnas.87.24.9868.

32. Chen JE, Glover GH. Functional Magnetic Resonance Imaging Methods. Neuropsychol Rev. 2015;25(3):289-313. doi: 10.1007/s11065-015-9294-9.

33. Pauling L, Coryell CD. The Magnetic Properties and Structure of Hemoglobin, Oxyhemoglobin and Carbonmonoxyhemoglobin. Proceedings of the National Academy of Sciences of the United States of America. 1936;22(4):210-6. doi: 10.1073/pnas.22.4.210.

34. Kim SG, Rostrup E, Larsson HB, Ogawa S, Paulson OB. Determination of relative CMRO2 from CBF and BOLD changes: significant increase of oxygen consumption rate

during visual stimulation. Magn Reson Med. 1999;41(6):1152-61. doi: 10.1002/(sici)1522-2594(199906)41:6<1152::aid-mrm11>3.0.co;2-t.

35. Faro SH, Mohamed FB. Functional MRI: Basic Principles and Clinical Applications. Springer; 2006.

36. Ogawa S, Menon RS, Tank DW, Kim SG, Merkle H, Ellermann JM, et al. Functional brain mapping by blood oxygenation level-dependent contrast magnetic resonance imaging. A comparison of signal characteristics with a biophysical model. Biophys J. 1993;64(3):803-12. doi: 10.1016/S0006-3495(93)81441-3.

37. Hoge RD, Atkinson J, Gill B, Crelier GR, Marrett S, Pike GB. Linear coupling between cerebral blood flow and oxygen consumption in activated human cortex. Proc Natl Acad Sci U S A. 1999;96(16):9403-8. doi: 10.1073/pnas.96.16.9403.

38. Cohen MS, Schmitt F. Echo planar imaging before and after fMRI: a personal history. NeuroImage. 2012;62(2):652-9. doi: 10.1016/j.neuroimage.2012.01.038.

39. Poustchi-Amin M, Mirowitz SA, Brown JJ, McKinstry RC, Li T. Principles and Applications of Echo-planar Imaging: A Review for the General Radiologist. 2001;21(3):767-79. doi: 10.1148/radiographics.21.3.g01ma23767.

40. Biswal B, Zerrin Yetkin F, Haughton VM, Hyde JS. Functional connectivity in the motor cortex of resting human brain using echo-planar mri. 1995;34(4):537-41. doi: 10.1002/mrm.1910340409.

41. Barkhof F, Haller S, Rombouts SARB. Resting-State Functional MR Imaging: A New Window to the Brain. 2014;272(1):29-49. doi: 10.1148/radiol.14132388.

42. Rubinov M, Sporns O. Complex network measures of brain connectivity: uses and interpretations. Neuroimage. 2010;52(3):1059-69. doi: 10.1016/j.neuroimage.2009.10.003.

43. McIntosh AR. Towards a network theory of cognition. Neural networks : the official journal of the International Neural Network Society. 2000;13(8-9):861-70. doi: 10.1016/s0893-6080(00)00059-9.

44. Friston KJ, Harrison L, Penny W. Dynamic causal modelling. Neuroimage. 2003;19(4):1273-302. doi: 10.1016/s1053-8119(03)00202-7.

45. Fox M, Raichle M. Spontaneous Fluctuations in Brain Activity Observed with Functional Magnetic Resonance Imaging. Nature reviews Neuroscience. 2007;8:700-11. doi: 10.1038/nrn2201.

46. Smith SM, Fox PT, Miller KL, Glahn DC, Fox PM, Mackay CE, et al. Correspondence of the brain's functional architecture during activation and rest. 2009;106(31):13040-5. doi: 10.1073/pnas.0905267106 %J Proceedings of the National Academy of Sciences.

47. Menon V, Uddin LQ. Saliency, switching, attention and control: a network model of insula function. Brain Struct Funct. 2010;214(5-6):655-67. doi: 10.1007/s00429-010-0262-0.

48. Lysen TS, Zonneveld HI, Muetzel RL, Ikram MA, Luik AI, Vernooij MW, et al. Sleep and resting-state functional magnetic resonance imaging connectivity in middle-aged adults and the elderly: A population-based study. 2020;29(5):e12999. doi: 10.1111/jsr.12999.

49. Klumpers LE, Cole DM, Khalili-Mahani N, Soeter RP, Te Beek ET, Rombouts SA, et al. Manipulating brain connectivity with δ^9-tetrahydrocannabinol: a pharmacological resting state FMRI study. Neuroimage. 2012;63(3):1701-11. doi: 10.1016/j.neuroimage.2012.07.051.

50. Khalili-Mahani N, Zoethout RM, Beckmann CF, Baerends E, de Kam ML, Soeter RP, et al. Effects of morphine and alcohol on functional brain connectivity during "resting state": a placebo-controlled crossover study in healthy young men. Hum Brain Mapp. 2012;33(5):1003-18. doi: 10.1002/hbm.21265.

51. Farras-Permanyer L, Mancho-Fora N, Montalà-Flaquer M, Bartrés-Faz D, Vaqué-Alcázar L, Peró-Cebollero M, et al. Age-related changes in resting-state functional connectivity in older adults. Neural Regen Res. 2019;14(9):1544-55. doi: 10.4103/1673-5374.255976.

52. Pievani M, de Haan W, Wu T, Seeley WW, Frisoni GB. Functional network disruption in the degenerative dementias. The Lancet Neurology. 2011;10(9):829-43. doi: 10.1016/s1474-4422(11)70158-2.

53. Galvin JE, Price JL, Yan Z, Morris JC, Sheline YI. Resting bold fMRI differentiates dementia with Lewy bodies vs Alzheimer disease. Neurology. 2011;76(21):1797-803. doi: 10.1212/WNL.0b013e31821ccc83.

54. Chhatwal JP, Schultz AP, Johnson K, Benzinger TL, Jack C, Jr., Ances BM, et al. Impaired default network functional connectivity in autosomal dominant Alzheimer disease. Neurology. 2013;81(8):736-44. doi: 10.1212/WNL.0b013e3182a1aafe.

55. Tian Y, Zalesky A. Characterizing the functional connectivity diversity of the insula cortex: Subregions, diversity curves and behavior. Neuroimage. 2018;183:716-33. doi: 10.1016/j.neuroimage.2018.08.055.

56. Gore JC. Principles and practice of functional MRI of the human brain. The Journal of clinical investigation. 2003;112(1):4-9. doi: 10.1172/JCI19010.

57. Hutton C, Bork A, Josephs O, Deichmann R, Ashburner J, Turner R. Image Distortion Correction in fMRI: A Quantitative Evaluation. NeuroImage. 2002;16(1):217-40. doi: https://doi.org/10.1006/nimg.2001.1054.

58. Jezzard P, Balaban RS. Correction for geometric distortion in echo planar images from B0 field variations. Magn Reson Med. 1995;34(1):65-73. doi: 10.1002/mrm.1910340111.

59. Maknojia S, Churchill NW, Schweizer TA, Graham SJ. Resting State fMRI: Going Through the Motions. Front Neurosci. 2019;13:825-. doi: 10.3389/fnins.2019.00825.

60. Zaitsev M, Akin B, LeVan P, Knowles BR. Prospective motion correction in functional MRI. Neuroimage. 2017;154:33-42. doi: 10.1016/j.neuroimage.2016.11.014.

61. Soares JM, Magalhães R, Moreira PS, Sousa A, Ganz E, Sampaio A, et al. A Hitchhiker's Guide to Functional Magnetic Resonance Imaging. 2016;10(515). doi: 10.3389/fnins.2016.00515.

62. Mikl M, Marecek R, Hluštík P, Pavlicova M, Drastich A, Chlebus P, et al. Effects of spatial smoothing on fMRI group inferences. Magnetic resonance imaging. 2008;26:490-503. doi: 10.1016/j.mri.2007.08.006.

63. Lv H, Wang Z, Tong E, Williams LM, Zaharchuk G, Zeineh M, et al. Resting-State Functional MRI: Everything That Nonexperts Have Always Wanted to Know. AJNR Am J Neuroradiol. 2018;39(8):1390-9. doi: 10.3174/ajnr.A5527.

64. Liu Y, Gao JH, Liotti M, Pu Y, Fox PT. Temporal dissociation of parallel processing in the human subcortical outputs. Nature. 1999;400(6742):364-7. doi: 10.1038/22547.

65. Li K, Guo L, Nie J, Li G, Liu T. Review of methods for functional brain connectivity detection using fMRI. Computerized medical imaging and graphics : the official journal of the Computerized Medical Imaging Society. 2009;33(2):131-9. doi: 10.1016/j.compmedimag.2008.10.011.

66. Iraji A, Calhoun VD, Wiseman NM, Davoodi-Bojd E, Avanaki MRN, Haacke EM, et al. The connectivity domain: Analyzing resting state fMRI data using feature-based data-driven and model-based methods. NeuroImage. 2016;134:494-507. doi: https://doi.org/10.1016/j.neuroimage.2016.04.006.

67. Snyder AZ, Raichle ME. A brief history of the resting state: the Washington University perspective. NeuroImage. 2012;62(2):902-10. doi: 10.1016/j.neuroimage.2012.01.044.

68. Lee MH, Smyser CD, Shimony JS. Resting-state fMRI: a review of methods and clinical applications. AJNR Am J Neuroradiol. 2013;34(10):1866-72. doi: 10.3174/ajnr.A3263.

69. Monti M. Statistical Analysis of fMRI Time-Series: A Critical Review of the GLM Approach. 2011;5(28). doi: 10.3389/fnhum.2011.00028.

70. Golestani AM, Goodyear BG. Regions of interest for resting-state fMRI analysis determined by inter-voxel cross-correlation. Neuroimage. 2011;56(1):246-51. doi: 10.1016/j.neuroimage.2011.02.038.

71. Zang Y-F, Jiang T, Lu Y, He Y, Tian L. Regional homogeneity approach to fMRI data analysis. NeuroImage. 2004;22:394-400. doi: 10.1016/j.neuroimage.2003.12.030.

72. Pedersen M, Curwood EK, Archer JS, Abbott DF, Jackson GD. Brain regions with abnormal network properties in severe epilepsy of Lennox-Gastaut phenotype: Multivariate analysis of task-free fMRI. Epilepsia. 2015;56(11):1767-73. doi: 10.1111/epi.13135.

73. Zou Q-H, Zhu C-Z, Yang Y, Zuo X-N, Long X-Y, Cao Q-J, et al. An improved approach to detection of amplitude of low-frequency fluctuation (ALFF) for resting-state fMRI: fractional ALFF. J Neurosci Methods. 2008;172(1):137-41. doi: 10.1016/j.jneumeth.2008.04.012.

74. Abdi H, Williams LJ. Principal component analysis. 2010;2(4):433-59. doi: 10.1002/wics.101.

75. Wold S, Esbensen K, Geladi P. Principal component analysis. Chemometrics and Intelligent Laboratory Systems. 1987;2(1):37-52. doi: https://doi.org/10.1016/0169-7439(87)80084-9.

76. Viviani R, Gron G, Spitzer M. Functional principal component analysis of fMRI data. Hum Brain Mapp. 2005;24(2):109-29. doi: 10.1002/hbm.20074.

77. Kiviniemi V, Kantola JH, Jauhiainen J, Hyvärinen A, Tervonen O. Independent component analysis of nondeterministic fMRI signal sources. Neuroimage. 2003;19(2 Pt 1):253-60. doi: 10.1016/s1053-8119(03)00097-1.

78. Calhoun V, Adali T, Hansen L, Larsen J, Pekar J. ICA of functional MRI data: an overview. 2003.

79. Beckmann CF. Modelling with independent components. Neuroimage. 2012;62(2):891-901. doi: 10.1016/j.neuroimage.2012.02.020.

80. Du Y, Allen EA, He H, Sui J, Wu L, Calhoun VD. Artifact removal in the context of group ICA: A comparison of single-subject and group approaches. Human brain mapping. 2016;37(3):1005-25. doi: 10.1002/hbm.23086.

81. Salvador R, Martínez A, Pomarol-Clotet E, Sarró S, Suckling J, Bullmore E. Frequency based mutual information measures between clusters of brain regions in functional magnetic resonance imaging. NeuroImage. 2007;35(1):83-8. doi: 10.1016/j.neuroimage.2006.12.001.

82. Mezer A, Yovel Y, Pasternak O, Gorfine T, Assaf Y. Cluster analysis of resting-state fMRI time series. Neuroimage. 2009;45(4):1117-25. doi: 10.1016/j.neuroimage.2008.12.015.

83. Craddock C, James A, Holtzheimer P, Hu X, Mayberg H. A whole brain fMRI atlas generated via spatially constrained spectral clustering. Human brain mapping. 2012;33:1914-28. doi: 10.1002/hbm.21333.

84. Shams SM, Afshin-Pour B, Soltanian-Zadeh H, Hossein-Zadeh GA, Strother SC. Automated iterative reclustering framework for determining hierarchical functional networks in resting state fMRI. Hum Brain Mapp. 2015;36(9):3303-22. doi: 10.1002/hbm.22839.

85. Watts DJ, Strogatz SH. Collective dynamics of 'small-world' networks. Nature. 1998;393(6684):440-2. doi: 10.1038/30918.

86. Bullmore E, Sporns O. Complex brain networks: Graph theoretical analysis of structural and functional systems. Nature reviews Neuroscience. 2009;10:186-98. doi: 10.1038/nrn2575.

87. Newman M. Newman MEJ.. Modularity and community structure in networks. Proc Natl Acad Sci USA 103: 8577-8582. Proceedings of the National Academy of Sciences of the United States of America. 2006;103:8577-82. doi: 10.1073/pnas.0601602103.

88. Di X, Biswal BB. Dynamic brain functional connectivity modulated by resting-state networks. Brain Struct Funct. 2015;220(1):37-46. doi: 10.1007/s00429-013-0634-3.

89. Tomasi D, Volkow ND. Functional connectivity density mapping. Proc Natl Acad Sci U S A. 2010;107(21):9885-90. doi: 10.1073/pnas.1001414107.

90. He Y, Chen ZJ, Evans AC. Small-world anatomical networks in the human brain revealed by cortical thickness from MRI. Cerebral cortex (New York, NY : 1991). 2007;17(10):2407-19. doi: 10.1093/cercor/bhl149.

91. Achard S, Salvador R, Whitcher B, Suckling J, Bullmore E. A resilient, low-frequency, small-world human brain functional network with highly connected association cortical hubs. The Journal of neuroscience : the official journal of the Society for Neuroscience. 2006;26(1):63-72. doi: 10.1523/jneurosci.3874-05.2006.

92. Smith SM, Jenkinson M, Woolrich MW, Beckmann CF, Behrens TE, Johansen-Berg H, et al. Advances in functional and structural MR image analysis and implementation as FSL. Neuroimage. 2004;23 Suppl 1:S208-19. doi: 10.1016/j.neuroimage.2004.07.051.

93. Smith SM. Fast robust automated brain extraction. 2002;17(3):143-55. doi: 10.1002/hbm.10062.

94. Lutkenhoff E, Rosenberg M, Chiang J, Zhang K, Pickard J, Owen A, et al. Optimized Brain Extraction for Pathological Brains (optiBET). PloS one. 2014;9:e115551. doi: 10.1371/journal.pone.0115551.

95. Jenkinson M, Bannister P, Brady M, Smith S. Improved Optimization for the Robust and Accurate Linear Registration and Motion Correction of Brain Images. NeuroImage. 2002;17(2):825-41. doi: https://doi.org/10.1006/nimg.2002.1132.

96. Greve DN, Fischl B. Accurate and robust brain image alignment using boundary-based registration. Neuroimage. 2009;48(1):63-72. doi: 10.1016/j.neuroimage.2009.06.060.

97. Beckmann CF, DeLuca M, Devlin JT, Smith SM. Investigations into resting-state connectivity using independent component analysis. Philosophical transactions of the Royal Society of London Series B, Biological sciences. 2005;360(1457):1001-13. doi: 10.1098/rstb.2005.1634.

98. Griffanti L, Salimi-Khorshidi G, Beckmann CF, Auerbach EJ, Douaud G, Sexton CE, et al. ICA-based artefact removal and accelerated fMRI acquisition for improved resting state network imaging. Neuroimage. 2014;95:232-47. doi: 10.1016/j.neuroimage.2014.03.034.

99. Branco P, Seixas D, Castro SL. Temporal reliability of ultra-high field resting-state MRI for single-subject sensorimotor and language mapping. NeuroImage. 2018;168:499-508. doi: https://doi.org/10.1016/j.neuroimage.2016.11.029.

100. Kondrak G, Marcu D, Knight K. Cognates can improve statistical translation models. Proceedings of the 2003 Conference of the North American Chapter of the Association for Computational Linguistics on Human Language Technology: companion volume of the Proceedings of HLT-NAACL 2003--short papers - Volume 2. Edmonton, Canada: Association for Computational Linguistics; 2003. p. 46–8.

101. Ptak R. The Frontoparietal Attention Network of the Human Brain:Action, Saliency, and a Priority Map of the Environment. 2012;18(5):502-15. doi: 10.1177/1073858411409051.

102. Kernbach JM, Yeo BTT, Smallwood J, Margulies DS, Thiebaut de Schotten M, Walter H, et al. Subspecialization within default mode nodes characterized in 10,000 UK Biobank participants. 2018;115(48):12295-300. doi: 10.1073/pnas.1804876115 %J Proceedings of the National Academy of Sciences.

103. Varangis E, Habeck CG, Razlighi QR, Stern Y. The Effect of Aging on Resting State Connectivity of Predefined Networks in the Brain. 2019;11(234). doi: 10.3389/fnagi.2019.00234.